RESEARCH AND PERSPECTIVES IN NEUROSCIENCES

Fondation Ipsen

J.-P. Changeux   A.R. Damasio
W. Singer   Y. Christen (Eds.)

Neurobiology
of Human Values

With 27 Figures and 2 Tables

 Springer

Changeux, Jean-Pierre, M.D, Ph.D.
Laboratoire de Neurobiologie
Moléculaire
Institut Pasteur
25, Rue du Docteur Roux
75015 Paris
France
e-mail: changeux@pasteur.fr

Damasio, Antonio R., M.D.
University of Iowa
College of Medicine
Iowa City
IA 52242
USA
e-mail: antonio-damasio@uiowa.edu

Singer, Wolf, Ph. D.
Max Planck Institute
for Brain Research
Deutschordenstr. 46
60528 Frankfurt/Main
Germany
e-mail: singer@mpih-frankfurt.mpg.de

Christen, Yves, Ph.D.
Fondation IPSEN
Pour la Recherche Thérapeutique
24, rue Erlanger
75781 Paris Cedex 16
France
e-mail: yves.christen@ipsen.com

ISSN 0945-6082
ISBN-10 3-540-26253-9 Springer-Verlag Berlin Heidelberg New York
ISBN-13 978-3-540-26253-4 Springer-Verlag Berlin Heidelberg New York

Library of Congress Control Number: 2005920457

Springer-Verlag is a part of Springer Science+Business Media

springeronline.com

Editor: Simon Rallison, Heidelberg
Desk Editor: Anne Clauss, Heidelberg
Production: PRO EDIT GmbH, 69126 Heidelberg, Germany
Cover design: design & production, 69121 Heidelberg, Germany
Typesetting: SDS, Leimen, Germany
Printed on acid-free paper   27/3150Re – 5 4 3 2 1 0

Contents

Contributors

Cela-Conde, Camilo J.
Laboratory of Human Systematics, University of Balearic Islands,
7071 Palma de Mallorca, Spain

Changeux, Jean-Pierre
Lab. Neurobiol. Moléculaire, Institut Pasteur, 25, Rue du Docteur Roux,
75015 Paris, France

Craighero, Laila
Dip. SBT, Sczione di Fisiologia Umana, via Fossato di Mortara, 17/19,
Università di Ferrara, 44100 Ferrara, Italy

Damasio, Hanna
University of Iowa College of Medicine, Department of Neurology,
Iowa City IA 52242, USA

Damasio, Antonio R.
University of Iowa College of Medicine, Department Neurology,
Iowa City IA 52242, USA

Davidson, Richard J.
Laboratory for Affective Neuroscience, University of Wisconsin,
1202 West Johnson Street, Madison, WI 53706, USA

Dehaene, Stanislas
INSERM Unit 562, "Cognitive Neuroimaging",
Service Hospitalier Frédéric Joliot, CEA, 4, Place de Général Leclerc,
91401 Orsay, France

de Waal, Frans B.M.
Emory University, Atlanta, Living Links, Yerkes Primate Center,
Emory University, 954 N. Gatewood Road, Atlanta GA 30322, USA

Greene, Joshua
Department of Psychology, Princeton University, Green Hall, Washington Road,
Princeton NJ 08544, USA

Houdé Olivier
Université Paris-5 et Centre Cyceron, CNRS et CEA,
UMR 6194 Groupe d'Imagerie Neurofontionnelle, Centre Cyceron,
Boulevard Henri Becquerel, 14074 Caen, France

Kahnemann, Daniel
Department of Psychology, Princeton University, Green Hall-Washington Road,
Princeton NJ 08544, USA

Rizzolatti, Giacomo
Dipartimento di Neuroscienze, Sczione di Fisiologia, via Volturno, 3,
Università di Parma, 43100, Parma, Italy

Singer, Wolf
Max Planck Institute for Brain Research, Deutschordenstr. 46,
60528 Frankfurt/Main, Germany

Sunstein, Cass R.
University of Chicago, Dept. of Political Science and the College,
111 East 60th Street, Chicago, IL 60637, USA

Introduction:
Neurobiology of Human Values

Yves Christen, Jean-Pierre Changeux, Antonio Damasio, Wolf Singer

Man has been pondering for centuries over the basis of his own ethical and aesthetic values. Until recent times, such issues were primarily fed by the thinking of philosophers, moralists and theologists, or by the findings of historians or sociologists relating to universality or variations in these values within various populations. Science – or to be more precise many scientists – has avoided this field of investigation within the confines of philosophy. In the name of seeking an objective truth, the scientific approach naturally avoid the field of normative truths, at the risk of appearing indifferent to the possible harmful consequences of its own discoveries. François Rabelais' famous dictum that "science sans conscience n'est que ruine de l'âme" ('science without conscience is but the ruin of the soul') testifies to this dichotomy between the world of values and that of dispassionate study of reality. This dichotomy is in large measure responsible for the mistrust of science displayed by some of the public, and the popular myth of the mad scientist, indifferent to the sufferings to which his discoveries give rise. We even sometimes hear it said that the purpose of moral philosophy might be to protect us from science! (Changeux in Changeux & Ricoeur 1998).

A revolution that is as much technological as conceptual

Beyond the temptation to stay away from the field of knowledge that is reserved for others – philosophers, politicians or religious figures – science may also have felt itself unconcerned by the study of human values for a simple heuristic reason, namely the lack of tools allowing objective study. For the same reason, researchers tended to avoid the study of feelings or consciousness until, over the past two decades, this became a focus of interest for many neuroscientists (Changeux 2002; Crick, 1994; A. Damasio 1999, 2003; Edelman, 1992; Llinas and Ribary 2001; W. Singer 2005a).

The rise of neuroscience and other disciplines makes a more objective and experimental approach possible. In particular, the imaging techniques used to display the living brain show us, for example, which areas of the brain are activated in association with moral or aesthetic ideas and attitudes. The field of neuroscience -brain imaging, as well as neurophysiology, neurology, anatomy, neuropsychology, etc.- has not been the only area to contribute new information. Ethology has enabled us to observe moral behavior – or behavior similar to ethics – in animals and also to apply an experimental approach to the sense of fairness and to other aspects of the biology of values. The respective observations and problems cover many disciplines, some of which fall in the domain of biology, while others are more a matter for social sciences such as sociology and economics. Some are directly linked to experience, while others are the subject of theoretical approaches, for example the modelling of the processes involved in biological evolution. We thought it would be beneficial to gather researchers from various disciplines to review the state of knowledge in this area and to discuss future prospects.

The distinction between the 'is' and the 'ought'

Our goal is not to create a new moral philosophy based on science. As quoted by Joshua Greene (2003) following David Hume (1739): "there is a sharp and crucial distinction between the 'is' of science and the 'ought' of ethics". In his "Positivist Catechism", Auguste Comte (1852) defended the opposite idea, that of the possible development of a science of moral philosophy that might open the way to a "positive moral philosophy", ensuring the triumph of sympathetic instincts at the expense of selfish people. But according to Cela-Conde (2005) and most others, we would make a logical mistake by trying to deduce values from facts.

We are also not willing to proclaim the superiority of science over other approaches to human values: it is simply a matter of noting that it offers the possibility of a new and objective form of knowledge. Whatever a person's convictions, this approach makes it easier to distinguish the line demarcating beliefs, opinions and convictions on the one hand and objective knowledge on the other.

Goodness

In recent years, several concrete observations have laid the foundations of the neuroscience of ethics. The personality changes and decline in moral sense suffered by Phineas Gage, the 19th century American railway worker who survived a metal bar passing through the front of his brain, provided evidence that social and moral functions are subserved by the frontal lobes (H. Damasio et al. 1994). Among the more recent observations pertinent to the neurobiology of goodness, we would consider in particular the implications of 1.) neuropsychological studies of cerebral lesions (particularly the ventro-medial prefrontal cortex, VMPFC) which demonstrated a loss of moral sense in subjects who became sociopathic or presented other behavioral deviations, irrespective of whether they benefited from normal development (Anderson et al. 1999; H. Damasio 2005), 2.) results from functional brain imaging of subjects responding to questionnaires designed to evaluate an ethical attitude (Greene et al. 2001; Greene 2005), 3.) the identification of neurons or neural networks involved in empathy (Rizzolati 2005) or of somatic markers that are themselves linked to emotions and thus cognition (A. Damasio 1994), 4.) experiments on primates – mainly chimpanzees and capuchin monkeys – performed to evaluate their sense of fairness (Brosnan & de Waal 2003; de Waal 2005), 5.) observations on animals living in their unconstrained habitat (de Waal 1996), 6.) investigations conducted by psychologists on the cognitive systems involved in making judgements and decisions – our knowledge of which may be increased by studying moral dilemmas and their contradictions

(Kahneman & Sustein 2005; Greene 2003) and finally, 7.) the brain responses to the acquired moral status of faces (T. Singer et al. 2004), that underline the importance of moral issues in social life.

One of the issues that emerges from this corpus of data is the naturality of moral behaviour, implying not only that there is a cerebral substrate for this behavior but also that it developed by evolutionary selection (Cela-Conde 2005). The latter idea corresponds closely to the views of Charles Darwin (1871), though curiously, not to those of all his successors. Even before ethology emerged as a discipline, developments in this field have moved in two opposite directions following the contributions of K. Lorenz, N. Tinbergen, E. Wilson and others. The thinking inherited from Darwin postulates that moral behaviour is natural in origin, and therefore both in man and animals results from the process of natural selection [the idea of a naturality of goodness was also supported by Kropotkine (1927)]. The thinking inherited from Thomas Huxley (1894), on the other hand, tends to regard moral philosophy as a supervenient element, which is peculiar to humans, and runs counter to genetic and evolutionary limitations [a trend well illustrated by Michael Ghiselin's famous and often repeated comment: "scratch an 'altruist', and watch a 'hypocrite' bleed" (1974)]. The multitude of books published over the past four decades on the subject of aggressiveness or conversely dealing with more peaceful attitudes, bears witness to this conflict. The work of Frans de Waal emphasises the positive behaviors such as reconciliation or empathy, which demonstrate the richness of social interactions. Indeed, depending on the circumstances, animals may be both aggressive and altruistic (de Waal 1989), but are certainly not systematically selfish (Sober & Wilson 1998). These studies provide arguments in favour of the Darwinian assumption of the naturality of social and moral behavior, and thus avoid having to explain how a species could have developed behavior that would be against its own nature!

Social and relational life is strongly linked to the emotions (A. Damasio 1994, 1999, 2003), which are themselves governed by an anatomical substrate that can be studied using brain imaging techniques. Richard Davidson (2005) found that activity in the left VMPFC is greatest in people who are very compassionate and caring. This work also emphasises the remarkable heterogeneity among individuals in their affective style and value. The diversity of both the subjects and the cultures in which they live obviously raises the question of ethical universalism or moral relativism, which has been extensively debated by Kant and other philosophers. Neuroscience can contribute to these issues (Changeux 1997; Changeux & Ricoeur 1998) – as well as other fields of course.

Beauty

The emotions also play a key role in aesthetics, though neuroscience has displayed less interest in this area. Nevertheless, J.-P. Changeux has pondered over the human brain's predisposition to invent or imagine creations (1994, 2005). In some ways, the artist may be regarded as a neuroscientist. For example, "artists have discovered which key contours must be perceived by the visual brain to identify the essential structure of an object. By studying the nature of lines used in line drawings, scientists may eventually gain access to this natural knowledge base" (Cavanagh 2005). Neurophysiologists such as S. Zeki (1993) have highlighted the responses of visual areas to drawings that he calls Mondrians. Cela-Conde et al. (2004) tried to localise aesthetic perception using magnetocencephalography. They found that the left prefrontal dorsolateral cortex was greatly activated when people perceived beautiful stimuli (either natural or artistic).

In a quite different field, prehistorians have shown the appearance of artistic expression, in the form of cave painting, in modern Homo sapiens with a brain similar to our own, during the Upper Paleolithic period in Europe, about 50 000 years ago according to many prehistorians (Klein 1999). Artists made beautiful painting in Grotte Chauvet 32 000 years ago (see Balter 1999), but the emergence of modern human behaviour and art could be older: two 77 000-year-old pieces of red ochre engraved with geometrical designs have been found at Blombos Cave on South Africa's south coast (Henshilwood et al. 2002).

Truth

Just as there is a physiology of goodness and beauty, there is also a physiology of truth, i.e. study of the neurobiological processes associated with understanding and acquiring knowledge (Changeux 2002). This comprises many aspects, both genetic and epigenetic, linked to the development of the brain, the mind and cultural acquisitions. The study of arithmetical abilities offers the benefit of allowing a certain quantification within the framework of neurospychological tests, and permits investigations performed on animals or on human societies that have not yet developed a sophisticated numerical system (Dehaene 2005). One enigmatic problem is that of understanding the mechanisms by which the brain knows that it holds the solution to the problem it submits to itself (W. Singer 2005b). On the other hand, Houdé (2005) provide the first insight into the cerebral basis of reasoning errors.

Neuroscience and human dignity

The Symposium organised by the Fondation Ipsen in Paris on 24 January 2005, summarized in the present proceedings, is one of the first events seeking to provide an overview of the neurobiology of human values. It goes without saying that it was not intented to arrive at final conclusions in an emerging field of science but to merely bring together researchers of various origins who are able to present original data or thoughts on this topic. This approach places a strong emphasis on the new achievements of neuroscience, but, it does by no means strive for hegemony. In fact this task is so vast that it calls not mainly for the expression of conflicting views but for the active collaboration among investigators of diverse disciplines.

As in other cases, some fear that the progress of scientific knowledge concerning attributes of the human being may have the effect that, what is magnificent when it lies hidden loses its sacred aura when revealed. The same might have been said about human love when progress was made in the field of neuroendocrinology. Our own belief, however, is that just as knowledge concerning hormones in no way diminishes the meaning or value of the passions involved in love, knowledge of the biological or neural substrate linked to truth, beauty or goodness will not impair the richness of these values in any way. In keeping with Jacques Monod's ethics of knowledge (1970), we consider that knowledge in itself constitutes a key value that constantly invites us to strive to find out more. History teaches us that degradation accompanies obscurantism more than does knowledge.

As J.Z. Young (1987) wrote: "It is time that people stopped talking about reductionism as if increased knowledge somehow subtracted from human dignity. On the contrary [it] adds greatly to understanding of our possibilities and limitations and hence ability to conduct ourselves wisely, and especially with the full respect for other human beings, and indeed for all life." Knowledge of the brain is already contributing to improving the human condition. Even though current psychiatry is still very at its beginning when it comes to understanding the sufferings involved in mental illness, it undoubtedly represents progress rather than an erosion of human dignity. Consequently, it is logical to hope that research into the origin and physiology of human values will be useful not only in terms of basic knowledge but perhaps also for helping mankind to cope with the turmoil that besets it. Neurological research might, for example, provide a better understanding of sociopathies and offer therapeutic solutions (H. Damasio 2005). Nevertheless, the decision to apply knowledge must be made by the society as a whole and not only by science. Here, we see that the progress in cognitive neuroscience raises a host of human issues. Our legal systems combine punishment with responsibility conceived as an absence of biological determinism. However neuroscientific progress indicates that some behaviours may occur as a result of biological factors presumably outside the control of individual responsibility. This consideration is at odds with the usual interpretation of moral and legal responsibility.

It is apparent that many questions linked to research in the field of neuroscience are now arising. The hope is that this book will help to formulate them more clearly rather than skirting them. The authors do not wish to launch a new moral philosophy, but simply to gather objective knowledge for reflection.

References

Anderson WW, Bechara A, Damasio H, Tranel D, Damasio AR (1999) Impairment of social and moral behavior related to early damage in the human prefrontal cortex. Nature Neurosci 2: 1032-1037

Balter M (1999) New light on oldest art. Science 283, 920-922.

Brosnan SF, de Waal FBM (2003) Monkeys reject unequal pay. Nature 425:297-299.

Cavanagh P (2005) The artist as neuroscientist. Nature 434:301-307.

Cela-Conde CJ, Marty G, Maestú F, Ortiz T, Munar E, Fernández A, Roca M, Rosselló J, Quesney F (2004) Activation of the prefrontal cortex in human aesthetic perception. Proc Ntl Acad Sci USA 101: 6321-6325.

Cela-Conde CJ (2005) Did evolution fix human values? In: Changeux J-P, Damasio A, Singer W, Christen Y (eds) Neurobiology of human values. Heidelberg: Springer Verlag, pp 11-15.

Changeux J-P (1994) Raison et plaisir. Paris : Odile Jacob.

Changeux J-P, ed. (1997) Une même éthique pour tous ? Paris : Odile Jacob.

Changeux J-P, Ricoeur P (1998) Ce qui nous fait penser. La nature et la règle. Paris : Odile Jacob [What makes us think ? Princeton University Press, 2000].

Changeux J-P (2002) L'homme de vérité. Paris : Odile Jacob [The physiology of truth. Harvard University Press, 2004].

Changeux J-P. (2002) Creation, art, and the brain. In: Changeux J-P, Damasio A, Singer W, Christen Y (eds) Neurobiology of human values. Heidelberg: Springer Verlag, pp 1-10.

Comte A (1852) Le catéchisme positiviste. Paris. New ed.: Paris, Garnier/Flammarion, 1966.

Crick F (1994) The astonishing hypothesis. The scientific search for the soul. New York: Charles Scribner's Sons.

Damasio A (1994) Descartes' error. New York: Putnam.

Damasio A (1999) The feeling of what happens. Body and emotion in the making of consciousness. New York: Harcourt Brace.

Damasio A (2003) Looking for Spinoza, Joy, Sorrow, and the Feeling Brain. New York: Harcourt.

Damasio A (2005) The neurobiological gounding of human values. In: Changeux J-P, Damasio A, Singer W, Christen Y (eds) Neurobiology of human values. Heidelberg: Springer Verlag, pp 47-56.

Damasio H, Grabowski T, Frank R, Galaburda AM, Damasio AR (1994) The return of Phineas Gage: the skull of a famous patient yield clues about the brain. Science 264:1102-1105.

Damasio H (2005) Disorders of social conduct following damage to prefrontal cortices In: Changeux J-P, Damasio A, Singer W, Christen Y (eds) Neurobiology of human values. Heidelberg: Springer Verlag, pp 37-46.

Darwin C (1871) The descent of man, and selection in relation to sex. London: John Murray.

Davidson R (2005) Neural substrates of affective style and value. In : Changeux J-P, Damasio A, Singer W, Christen Y (eds) Neurobiology of human values. Heidelberg: Springer Verlag, pp.

Dehaene S (2005) How a primate brain comes to know some mathematical truths. In: Changeux J-P, Damasio A, Singer W, Christen Y (eds) Neurobiology of human values. Heidelberg: Springer Verlag, pp.

de Waal FBM (1989) Peacemaking among primates. Cambridge, MA: Harvard University Press.

de Waal FBM (1996) Good natured: the origins of right and wrong in humans and other animals. Cambridge: Harvard University Press.

de Waal FBM (2005): Homo homini lupus? Morality, the social instincts, and our fellow primates. In: Changeux J-P, Damasio A, Singer W, Christen Y (eds) Neurobiology of human values. Heidelberg: Springer Verlag, pp 17-35.

Edelman GM (1992) Bright air, brilliant fire: on the matter of mind. New York: Basic Books.

Ghiselin M (1974) The economy of nature and the evolution of sex. Berkeley: University of California Press.

Greene JD, Sommerville RB, Nystrom L E, Darley JM, Cohen JD (2001) An fMRI investigation of emotional engagement in moral judgement. Science 293:2105-2108.

Greene J (2003) From neural 'is' to moral 'ought': what are the moral implications of neuroscientific moral psychology. Nature Neurosci 4:847-850.

Greene J (2005) Emotion and cognition in moral judgment: evidence from neuroimaging.In: Changeux J-P, Damasio A, Singer W, Christen Y (eds) Neurobiology of human values. Heidelberg: Springer Verlag, pp 57-66.

Henshilwood CS, d'Errico F, Yates R, Jacobs Z, Tribolo C, Duller GAT, Mercier N, Sealy JC, Valladas H, Watts I, Wintle AG (2002) Emergence of modern human behaviour: middle stone age engravings from South Africa. Science 295:1278-1280.

Houdé O (2005) Cerebral basis of human errors. In: Changeux J-P, Damasio A, Singer W, Christen Y (eds) Neurobiology of human values. Heidelberg: Springer Verlag, pp.

Hume D (1739) A treatise of human nature. Selby-Bigge LA, Nidditch PH (eds) Oxford: Clarendon, 1978.

Huxley TH (1894) Evolution and ethics. New ed.: New York, Appleton, 1898.

Kahneman D, Sustein CR (2005) Cognitive psychology of neural intuitions. In: Changeux J-P, Damasio A, Singer W, Christen Y (eds) Neurobiology of human values. Heidelberg: Springer Verlag, pp.

Klein RG (1999) The human career: human biological and cultural origins. Chicago: University of Chicago Press.

Kropotkine P (1927) L'éthique. [New ed.: Paris: Stock, 1979].

Llinas R, Ribary U (2001) Consciousness and the brain. The thalamocortical dialogue in health and disease. Ann N Y Acad Sci 929:166-175.

Monod J (1970) Le hazard et la nécessité. Paris : Le Seuil.

Rizzolati G (2005) Mirror neurons: a neurological approach to empathy. In: Changeux J-P, Damasio A, Singer W, Christen Y (eds) Neurobiology of human values. Heidelberg: Springer Verlag, pp.

Singer T, Kiebel SJ, Winston JS, Dolan RJ, Frith CD (2004) Brain responses to the acquired moral status of faces. Neuron 41:653-662.

Singer W (2005a) Large-scale temporal coordination of cortical activity as prerequisite for conscious experience. In: Nirit S (ed.) Companion to Consciousness. Malden, MA: Blackwell, in press.

Singer W (2005b) How does the brain know when it is right? In: Changeux J-P, Damasio A, Singer W, Christen Y (eds) Neurobiology of human values. Heidelberg: Springer Verlag, pp

Sober E, Wilson DS (1998) Unto others: the evolution and psychology of unselfish behaviour. Cambridge, MA: Harvard University Press.

Young JZ (1987) Philosophy and the brain. Oxford: Oxford University Press.

Zeki S (1993) A vision of the brain. Oxford: Blackwell.

Creation, Art, and the Brain*

Jean-Pierre Changeux[1]

Visitors to the Imperial Collections at the National Museum of Taipei are struck upon entering the exhibit by a peculiar set of objects: oracle bones from the Shang Dynasty. These fragments of tortoise shell feature unusual fractures, mostly randomly distributed, and resulting from the insertion of white-hot firebrands into cavities that are pierced along the internal wall of the bone. With respect to these fissures, texts written in the oldest-known Chinese characters attribute meaning to each fracture line: the answer (auspicious or ominous) to a set of questions asked by a fortune-teller (military success, harvest, illness, dreams, and so on) and, by logical extension, the supposed verification of the prediction.

These first written traces testify to an activity that is fundamental to the human brain: that of giving meaning to a physical makeup (in this case, randomly distributed line patterns) with the goal of mastering the evolution of time in a world that escapes us. The human brain proceeds in a direction opposite to the "input-output" mode long proposed by cybernetics. It always projects onto the world, in a spontaneous and internally generated fashion, mental representations that it tries to test against an external reality that is intrinsically devoid of meaning. This projection, as a generator of mental forms, represents an essential predisposition of the human brain toward creation (Changeux, 1983, 2004; Berthoz, 2002).

Scientists are cautious in discussing a subject as difficult as "creation." Experimental data available to the neurobiologist remain fragmentary and limited. In the absence of human data, animal models are useful, but insufficient. In addition, current scientific methods do not yet allow us objectively to reconstruct the life experience of the creator and his imaginary world. My discussion will remain hypothetical.

The Human Brain's Predisposition toward Creation

Is it plausible that a group of nerve cells and their axonal and dendritic extensions, in which electrical and chemical signals travel, could be at the origin of

* Copyright © by the President and Fellows of Harvard College. All rights reserved. Printed in the United States of America.

[1] Lab. Neurobiol. Moléculaire, Institut Pasteur, 25, Rue du Docteur Roux, 75015 Paris, France; e-mail: changeux@pasteur.fr

a function as seemingly elusive as creation? At this stage, I would only say that there is no physical reason why not (Changeux, 1983, 2004; Edelman, 1992). The human brain is made up of approximately 100 billion neurons, and each neuron has an average of 10,000 discontinuous contacts (or synapses), with its multiple connections. Thus there are on the order of 10^{15} synapses in the human encephalon, or approximately 600 million synapses/mm^3. The number of possible combinations between neurons reaches astronomic powers of 10, followed by millions of zeros, on the same order of magnitude as the number of positively charged particles in the universe. In addition, the efficiency of these synaptic contacts varies as a function of experience. No numerical limit can legitimately be placed on the functional combinations of the neuronal network.

Moreover, connections in the neuronal network are neither totally random, like a gas, nor completely regular, like a crystal. They are organized. Relationships of convergence – of many axons on one neuron – and of divergence – of the same axon on several neurons – succeed in bringing together local specialization and global integration within the cerebral network. Finally, the brain's organizational pattern, stable from one generation to the next, results from an essentially genetic, biological evolution that is several hundred million years old. During this evolution, according to Herbert Spencer's description, "coordination between sensory and motor groups developed and, were made even more complex by the introduction of new elements combined in various ways to make up spheres of increasingly vast extent that could become mentally present" with, continues Hughlings Jackson, "passage from the simplest toward the most complex, that is from the well-organized, inferior center, to the superior, comprehends. With complete understanding of the story, a new cortical region comes into play: the prefrontal cortex. It is known that its relative surface increases in size exponentially from the monkey to the human (Fuster, 1980). It is also known that damage to this area results in serious behavioral problems, that, according to the specific site of the lesion, affect attention, reasoning, or mental planning. The prefrontal cortex contributes to mental synthesis, to organizing thoughts into intentions. It helps to build mental representations that relate to social interaction. It plays a central role in attributing mental states to others, in the recognition of one's own "singular otherness," to use the term of the French philosopher Emmanuel Lévinas, and thus contributes to one of the central abilities of the brain in the appreciation of aesthetic qualities of artwork and in moral judgment: empathy.

The concept of empathy calls forth an entire domain of the creative brain to which I have not yet alluded: that of the emotions (Damasio, 1995). Ancient and primitive, some would say "animal" (because of its role in vital survival functions), the limbic system plays an essential part in assessing danger as well as pleasure and in shaping the subject's physical and emotional landscape. It is linked to neuronal systems of the brain stem that underlie self-stimulation among rats. Intricately connected to the frontal cortex, the limbic system marks out emotional reference points to guide other systems that act upon the world. Because of the limbic system, reason and pleasure are allowed to live in harmony in the brain of the creator.

Creation and Mental "Darwinism"

In defense of the "genetic critique" of the literary creation process, illustrated in manuscripts by Gustave Flaubert or Francis Ponge, Almuth Grésillon (1992) emphasizes that "against the straight highway, the inexorable march toward resolution and outcome, the teleology of the straight line, we find opposite metaphors that indicate a more winding path: forks, turnoffs, wrong turns, unbeaten paths, detours, crossroads, backtracks, dead ends, accidents, and false starts." According to Claude Simon, "the writer progresses laboriously, gropes around blindly, gets bogged clown, and starts up again." Matisse compares the path of his pencil to "the gesture of a man feeling his way forward in the dark." In many respects, the memory of Jacques-Louis Lions's (Franck, 1995) black painting, preserving what Laurent Schwartz, a French mathematician, refers to as the multiple "sharp lines" and "zigzags," is more informative about the mechanisms of mathematical creation than is a report published in the Proceedings of the Academy of Sciences. The process of creation proceeds along a slow road of trial and error.

Artists' and scientists' brains have not yet been imaged in the PET scanner to record the creative process in an objective manner. We are restricted to a "diet" of data that confines us to schematic models of theorization. The proposed hypothesis (Changeux, 1983, 2004; Changeux and Dehaene, 1989; Dehaene et al 1998, 2003) is that creation proceeds, in psychological time, according to an evolving process of trial and error. This process encompasses biological evolution and connectional epigenesis within a framework of "mental Darwinism", and according to a projective style. During periods of "preparation" and "incubation" – in the words of Jacques Hadamard, a famous French mathematician – fragmentary images of sketches of mental representations or prerepresentations (Changeux and Dehaene, 1989) pop up spontaneously and transiently in the brain. A combinatory function, a kind of "mental handiwork," associates and dissociates objects from memory and current perceptions, until, in this competition, one "recombination" wins out. According to Hadamard, this is enlightenment or, more generally, the selection of a particular representation. This meaningful object will generally serve as a step in the halting, constantly renewed process of mathematical invention, biologic modeling, or in the early design of a work of art. This work of the imagination will sooner or later engage a selection-by- evaluation mechanism that, seemingly, brings into play the limbic system and its outposts, already mentioned in the context of emotion. The limbic system, flooding the synapses of the relevant neuronal group with neurotransmitters, thus modifies their efficiency and, consequently, produces a material trace in long-term memory.

As one might expect, the space in which these mental images "struggle for life," as Taine wrote, is not perfectly identified. The brain analog of Claude Bernard's "interior milieu," this conscious milieu, a "simulation space" of virtual activities, provides a meeting place for current perceptions of the outside world, for representations of "self," and for memories of past experiences, somatically marked with emotional valence. Also contributing to these deliberations are many internalized social rules, as well as innate forces that stem from biological evolution and whose influence is often beyond the scope of the subject's perception. The physical basis of this "conscious milieu" has been postulated to consist of a global network composed of neurons with long range axons which broadcast signals to and from multiple areas of the cerebral cortex yielding subjective experience of being conscious. Since these pyramidal neurons are particularly abondant in layers two and three of the association areas such "global workspace" seemingly includes the frontal cortex on tops of "automatic" specialized cortical processors (Dehaene et al 1998, 2003) the limbic system, and the brain stem, among many other brain areas.

Touchstones of a "chemical" definition of the conscious milieu, psychotropic drugs, such as mescaline, bind selectively with certain categories of receptors and neurotransmitters (in this case catecholamines). The Huichol Indians take mescalinie as part of an annual ritual during their pilgrimages to Wirikuta. They then often experience multicolored hallucinations, with strong geometric components of remarkable symmetry. Once they have returned home, they set these fleeting images to wool canvases that are supposed to represent their journey to "the beyond." Could it be that these colored and regular images reflect the neural geometry of the color areas of V4? Do they engage the same receptor, as when similar shapes emerge, albeit in a voluntary way, in the minds of artists such as Kandinsky, Manessier, or Messiaen, who "heard" colors when he was creating?

During the "mental experience" of creative work, multiple evolutions overlie and connect with each other. Beginning with the first "confused shapes" (Leonardo da Vinci), "randomly born bits" on which, according to the French poet and essayist Yves Bonnefoy, "the imaginary rests," one can distinguish the "first idea," a still-crude outline that is nevertheless so importantly connected to the subject's own definition. Then this mental object, sifted through the filter of reason, becomes reality on paper by multiple "external drawings" of the graphic sketch, or disegno. Representations from the "internal circuit" give way to precise movements of the hand and fingers (often acquired after a long period of practicing one's art) that finally direct the pen or the paintbrush. A cascade of activity in premotor, then motor areas extends the internal, implicit evolution of mental creation.

The neuroscientist Apostolos Georgopoulos (Georgopoulos, Taira, and Lukashin, 1993) has deciphered, using monkey models, the activity of motor neuron populations responsible for guiding hand movements. The average vector of the recorded neural activities, in frequency and in direction, coincides remarkably well with the orientation vector for finger movement, especially when the animal is drawing a spiral (Schwartz, 1994). An "evolutionary dialog" takes shape between the brain of the creator and the work being created, "adjustments of the eye with reasoning," which accompany corrections, cross-outs, and restarts of the work in progress.

Individual and Cultural Histories of Creation

Artistic and scientific creation are part of an individual history which itself stems from an anterior historical evolution. Thus Hélène Seckel (1988) would place La vision de Saint-Jean, (Saint John's Vision) by El Greco at the origins of Picasso's Demoiselles d'Avignon, followed by Les baigneurs (The Bathers) by Cézanne, the Bain turc (Turkish Bath) by d'Ingres, and, among Picasso's own works, Le harem de 1906 (The Harem of 1906) and Deux femmes nues (Two Nudes) from the same year, with an obvious influence of African art, and finally several sketches for the painting itself.

In the same way, every scientist, as much as he may think himself a genius, knows what he owes to the cultural environment in which he took his first steps, to the discoveries of his predecessors, to unstructured meetings with colleagues, to accidental observations, to references gleaned here and new techniques borrowed there. In this elusive network of unforeseen interactions, where contingency and rationalization mix, "cognitive crystallizations" take shape. Their eventual validation and final acceptance will occur when universally aimed critical debates are held. Historians of science (Creager and Gaudillière, 1996) have conducted a sophisticated analysis of the inherent difficulties of untangling such a web. To the problem of conceptual content is added that, no less important, of claims of "priority" or "credit" (essential to the psychology of even the most eminent scientists), which mask the processes that are intrinsic to the evolution of knowledge (or even lead us astray on the subject). Thus research on the history of science prefers to count not so much on unreliable written or oral autobiographical accounts, but rather on in-depth examinations of experimental note books, on exchanges of correspondence, and on other testimony that is more neutral from the point of view of the personal visibility of the researchers (Creager and Gaudillière, 1996).

The historic evolution of artistic and scientific representations does not, however, follow as obscure and random a path as it may seem. It engages a rather severe selection mechanism following specific rules – mental tools – that creators have imposed on themselves from the beginning of time, and that they themselves have made evolve throughout the course of history. "I love the rule that corrects emotion," notes the French cubist Georges Braques. And Louise Bourgeois (1995) continues: "My sculptures are like algebraic equations with variables. The equations must be tested."

A first, very general, rule is the coherence between the parts and the whole. For the mathematician, the coherence of a new mathematical object with mental objects that already exist in his brain marks a true "invention." In the realm of art, Alberti (1970 [1435]), taking after the Stoics, sets down as an absolute law the principle of harmonia or consensus partium, which occurs "if all the elements, whether they be dimension, function, type, color, and other similar things, converge upon a single and same beauty." Matisse seems to concur with Alberti when he writes: "I take from Nature what I need, I carefully combine all the effects, I balance them in description and in color, and this condensation in which everything harmonizes is never reached on the first attempt." The successive steps, photographed by Matisse himself, that led to the creation of Grand nu allongé (Large Reclining Nude) (1935), displayed at the Baltimore Museum, testify to this tireless search, for "the harmony of the whole" when creating a work intended to become a masterpiece.

The neuronal counterpart to what Diderot calls "the apperception of ratios" still remains very enigmatic. I will first discuss a series of experiments that examine "illusory contours." These occur, for example, when a subject perceives a continuous luminous bar connecting notches cut into two darker elements, distinguishable when the notches remain open on a light-colored background. Peterhans and Von der Heydt (1989) have shown that signature neural activity, in this case in area V2, characterizes the occurrence of the illusory contour of the luminous bar. The brain constructs a coherent representation of the world, even if only virtual. This goes hand in hand with an even more fundamental perception of the invariability of the world that surrounds us. When we voluntarily change the position of our head or the orientation of our gaze, we feel like the outside world is immobile and that we are moving within it. This perception poses a serious problem of reference coordinates for our brain. Droulez and Berthoz (1988) suggest that the brain constructs an internal model of the environment, which it updates continuously by subjecting our sensory representations to the overriding input from motor-based controls responsible for the orientation of our head and gaze. The connections linking temporal and frontal areas may be said to play a role in the synthesis of that mental object, unique and coherent, to which artist and scientist may well refer in their search for harmony.

In this context, several series of experiments refer more specifically to symmetry, a concept dear to the hearts of mathematicians and architects, and one found abundantly in Islamic art and Buddhist mandalas. Among many types of swallows (Moller, 1992) and earwigs, it bas been shown that females prefer males who are symmetric. The evolutionarily based reasons that are given, valid for superior vertebrates as well, vary from the "honest ad effect" to the universal necessity to recognize objects whatever their orientation in the outside world. Among humans, the attractiveness of a face does not depend only on its symmetry. Perrett (Perrett, May, and Yoshikawa, 1994), using composite digitized photographs, has shown that, contrary to popular belief, the "average" face is not the most attractive. Japanese, like Caucasians, in fact prefer a composite face constructed from facial features individually recognized as being the most seductive, the exaggeration of shape differences compared with the norm making the composite that much more attractive across cultures. There is a search for "individuality" amid a basic symmetric organization. The ultimate coherence of perfect symmetry yields to the search for quasi-symmetry, for the baroque style that we see reappear throughout the history of art.

One rule on which all agree in science as in art is that of novelty, which excludes the familiar, things already seen or heard before. A powerful brain device comes into play in detecting the novel, the unpredicted, a device essential to the survival of the organism. Pavlov and then Sokolov have shown how an animal subjected to a repetitive stimulus, let's say a sound, ends up no longer responding to that stimulus. A "desensitization" occurs. However, as soon as the stimulus changes, the organism produces an orientation response toward the source, by its gaze and its posture. Thus this important brain device comes into play not only in the detection of novelty but, consequently, in its production as well. Neuroimaging observations by Posner reveal, along these same lines, that when a subject becomes familiar with a list of nouns that he has read several times over, brain activity gradually turns off. But as soon as a new list is presented, the frontal (cognitive) cortex and Wernicke's area (language) light up on the lateral surface, as well as the cingulate cortex (limbic) on the medial surface. No doubt reading a poem by René Char, a melody by Poulenc, or a Proceedings paper by Louis Pasteur would maintain a permanent level of high blood flow in the listener, whose brain could be said to "echo" that of the creator.

Coherence and novelty are insufficient to select mental representations that deserve the attributes of "true" or "beautiful." The rule of reality testing (adéquation au réel) had united science and art for centuries, until romanticism created an irremediable rift. In the natural sciences, experimentation rules. A very attractive mathematical model can, from one day to the next, be relegated to the throwaways of history because of its incompatibility with external reality. This is what happened to Pauling's triple-helix DNA model after the presentation and subsequent validation of Watson and Crick's double-helix model. In the natural sciences, invention cannot stray from the narrow and demanding path of discovery of a preexisting physical reality and the mastery of the rules that govern it.

"Mental tools" that are produced by the scientist's brain, theoretic models, as adequate as they may seem, always remain limited and cannot pretend to exhaust the reality of the world; at most, they can only define its "horizons." At the same time, the validation of a model cannot be accomplished by a single human being: it necessitates collective, open, critical debate, an exchange of "representations," as already mentioned, between scientists' brains. Whatever may be said about it, the application of the rule of "reality testing," which makes one accountable to the principles of objectivity and universality in scientific knowledge, leads to indisputable progress in the field of human knowledge.

The idea of progress in art appears, to the contrary, to be absurd. Certainly, from the beginning of time, the artist has referred to the natural world. For example, the composer André Jolivet, in his work, uses the natural rhythms of the heart, of breathing, and of sex. The painter Mondrian uses lines, basic shapes, and colors that, as we have seen, are underpinned by fundamental neural mechanisms of visual perception. However, is Stravinsky's Le sacre du printemps a fundamental improvement over Monteverdi's Orfeo? No, because if art evolves and renews itself, the artist has long since broken his links with the objective description of nature and its reconstruction into an "ideal beauty." According to Matisse's terms, the artist "organizes his sensations," "releases his emotions" in order to solicit those of the viewer (or the listener) and touch his subjectivity.

The artist thus seems "freer" than the scientist to define the rules of his creation, and the rules in fact do vary from one artist to another. These rules actually characterize an artist's style. For example, for Matisse, "what I dream of is an art of balance, of purity, of tranquility, without any worrisome or troublesome subject, an art that would be, for any cerebral worker... something that calms and soothes the brain, like a comfortable armchair that relaxes and cures one from physical tiredness." Conversely, Marcel Duchamp announces the "death of art" by invoking humor and derision. For Ben (Vautier), a painter based in Nice, France, one of the functions of art is to distract, to capture the eye. The French painter François Morellet, with his "Geome-trees," attempts "a freak marriage between Pythagoras and Buffon" by evoking, through strictly abstract shapes, trees, curly locks of hair, and copulations. In the musical domain, Pierre Boulez searches for "slippages avec overlays" ("les glissements effectués sur leurs superpositions", the contrast of tones, disturbances, and accidents in "a universe where consonance and dissonance are abolished." Among many contemporary artists, a return to the origins of creation, to its brain-based mechanisms, gives rise to a pictorial fiction, to some extent moving against the grain of time, leaving the spectator free to complete the creative process by the greatest number of possible routes. The random generator of the diversity of mental representations is in a sense re-created with the automatic writing of the surrealists, the "dionysian dance" of Pollock's "dripping," the "conscious awkwardness" of Cy Twombly's scribbles, Dubuffet's "psychosites," Carcassonne's "neurographies," the production of Frank Stella's "images [that] lead into each other while always continuing to distort themselves," and that are like as many "neuromotor pantomimes" (Jean Clair) of internal experience.

In the course of history, creators have thus invented "rules" that characterize their styles. But these rules are not set in stone, and they evolve. The perspective in painting, like the temperament in music, tries to seem "right" or to "ring true," but in fact deceives the eye and ear. The internalization of these rules implies a long and rather irreversible apprenticeship, similar to the learning of a language. The study of babies' listening strategies by Trehub and Trainor (1994) shows that they treat relations of pitch and the passage of time in the same way as adults. Babies spontaneously prefer "salient" musical form with perfect fifth and keynote intervals. However, at the age of six months, Western and Javanese infants can detect wrong notes in melodies based in the Western major scale and the Javanese pelog scale, while adults are able only to recognize wrong notes from the musical scale of the culture in which they were raised. This drop in performance, which can be thought of as a "perceptual calibration" by selection of innate patterns (Changeux, 1983; Changeux and Danchin, 1976), severely limits the widespread notion of our perceptual system's virtually limitless plasticity. This process, in accordance with the theory of epigenesis by selective stabilization, finds itself in opposition to the hypothesis of composers such as Boulez or Schoenberg, according to whom adult listeners may acquire all imaginable musical structures with the same ease with which they were able to perceive or acquire conventional musical structure as children. In this study, Trehub and Trainor emphasize the double constraint of innate dispositions, of "neural forms" proper to the species, and of epigenetic selection based in individual and cultural history.

In wanting to return to the very origins of creation, wouldn't art be returning to that "natural chaos," those "confused shapes" that it is, to the contrary, supposed to organize? No, because, in my opinion, art possesses an essential function at the societal level. It plays a part in the production of relevant (Sperber and Wilson, 1989) forms that respond to the virtual (but nevertheless real) landscape of human societies' "subjective expectations" and because of this it contributes to the religere, "careful reflection," of individuals in the social group.

Louise Bourgeois (1995) writes that the sculpture Silent Black Woman (Silencieuse femme noire) represents the protest marches of the 1960s. And she continues: "There are many types of tensions, but the one I am trying to resolve, to soothe, is social, tension." Here, we approach the "noble ideas" (belles idées) of the exemplum that Nicolas Poussin invites us to illustrate by our own actions. The work of art, through its capacity to awaken and make aware, thus allows one to pass on an ethical message. And in spite of this, one may wonder why dreaded human blindness, which caused the deaths of the Children of Bethel, as Laurent de la Hire bas shown us, still prevails over our planet? Does not the irresponsibility of certain scientists who, with "professional conscience," reiterate in horror – even with good intentions – the tragedy of the Children of Bethel, deserve to be unmasked under any circumstances? Cannot art – news photography, for example – cry out loudly enough to prevent humanity from such straying off the right path of science?

I began this chapter with the image of a fortune-telling tortoise shell from the Shang Dynasty, emphasizing how humans, in the beginning of civilization, projected meaning onto that which had none. With the assets of universal scientific knowledge, human beings should make a commitment to use the creative faculties they possess in their brains to give meaning to that which calls out for it the most: humanity itself. It is our responsibility to urgently invent an "ethical model," which breaks with the violence, the intolerance, the crimes of our cultural past, and ensures more efficiently survival and well-being for all human lives.

References

Alberti LB (1970) On painting (Della pittura). Translated with introduction and notes by John R. Spencer. New Haven: Yale University Press [1435]

Berthoz A (2002) The brain's sense of movement. Cambridge: Harvard University Press

Boulez P (1971) Boulez on music today (Penser la musique aujourd'hui). London: Faber

Bourgeois L (1995) Catalogue d'exposition. Paris: Musée d'art moderne de la ville de Paris

Changeux JP (1983). L'homme neuronal. Paris: Fayard

Changeux JP (1994) Raison et plaisir. Paris: Odile Jacob

Changeux JP (2004) The physiology of truth. Cambridge: Harvard University Press

Changeux JP, Connes A (1989) Matière a pensée. Paris: Odile Jacob

Changeux JP, Danchin A (1976) The selective stabilization of developing synapses: A plausible mechanism for the specification of neuronal networks. Nature 264: 705–712

Changeux JP, Dehaene S (1989) Neuronal models of cognitive functions. Cognition 33: 63–109

Changeux JP, Chavaillon J, eds. (1995) Origins of the human brain. Oxford: Clarendon

Creager ANH., Gaudillière JP (1996). Meanings in search of experiments and vice-versa: The invention of "allosteric regulation" in Paris and Berkeley (1959-1967). Historical Studies in the Physical and Biological Sciences 27: 1–89

Damasio A (1995) L'erreur de Descartes. Paris: Odile Jacob

Debru C (1983) L'esprit des protéines. Paris: Hermann

Dehaene S, Kerszberg M, Changeux JP (1998) A neuronal model of a global workspace in effortful cognitive tasks. Proc Natl Acad Sci USA 95: 14529–34

Dehaene S, Sergent C, Changeux JP (2003) A neuronal network model linking subjective reports and objective physiological data during conscious perception. Proc Natl Acad Sci USA 100: 8520–5

Dretske F (1995) Naturalizing the mind. Cambridge: MIT Press

Droulez J, Berthoz A (1988) Servo-controlled conservative versus topological (projective) modes of sensory motor control. In Disorders of posture and gait, ed. W. Bles and T. Brandt, eds. Amsterdam: Elsevier 83–97

Edelman G (1992) Biologie de la conscience. Paris: Odile Jacob

Fellemann D, Van Essen D (1991) Distributed hierarchical processing in primate visual cortex. Cerebral Cortex 1 : 1–47

Franck M (1995) Collège de France: Figures et travaux. Paris: Imprimerie nationale

Fuster J (1980) The prefrontal cortex. New York: Raven Press

Georgopoulos A, Taira M, Lukashin A (1993) Cognitive neurophysiology of the motor cortex. Science 260: 47–51

Grésillon A (1992) Ralentir travaux. Genesis 1: 9–31

Hebb D (1949) The organization of behavior. New York: Wiley

Lawrence P (1992) The making of a fly. Oxford: Blackwell

Le Bihan D, Turner R, Zeffiro T, Cuénod C, Jezzard P, Bonnerot V (1993) Activation of human primary visual cortex during visual recall: A magnetic resonance imaging study. Proc Natl Acad Sci 9 USA 0: 11802–11805

MacAdams S, Bigand E, eds. (1994) Penser les sons. Paris: PUF

Mazoyer BM, Tzourio N, Frak V, Syrota A, Murayama N, Levrier O, Salomon G, Dehaene S, Cohen L, Mehler J (1993) The cortical representations of speech. J Cognitive Neurosci 5: 467–469

Miyashita Y, Chang H (1988) Neuronal correlate of pictorial short term memory in the primate temporal cortex. Nature 331: 68–70

Moller AP (1992) Female swallow preference for symmetrical male sexual ornaments. Nature, 357 : 238–240

Monod J (1970) Le hasard et le nécessité. Paris: Editions du Seuil

Movshon JA, Blakemore C (1974) Functional reinnervation in kitten visual cortex. Nature 251: 504–505

Perrett D, May K, Yoshikawa S (1994). Facial shape and judgements of female attractiveness. Nature 368: 239–242

Perrett D, Oram MW (1993) Neurophysiology of shape processing. Image and Vision Computing 11: 317–330

Perrett D, Rolls ET, Caan N (1982) Neurons responsive to faces in the monkey temporal cortex. Experimental Brain Res 47: 329–342

Peterhans E, Von der Heydt R (1989) Subjective contours: Bridging the gap between psychophysics and physiology. Trends Neurosci 14: 112–118

Posner M, Raichle M (1994) Images of mind. New York: Scientific American Library

Pusher M (1993) Seeing the mind. Science 262: 673–674

Schwartz A (1994) Direct cortical representations of drawing. Science 265 : 540–542

Seckel H (1988) Les demoiselles d'Avignon. Catalogue d'exposition. Paris: Réunion des musées nationaux

Sereno M, Dale A, Peppas T, Kwong K, Belliveau J, Brady T, Rosen B, Tootel R (1995) Borders of multiple visual areas in humans revealed by functional magnetic resonance imaging. Science 268 : 889–893

Sperber D, Wilson D (1989) La pertinence. Paris: Editions de Minuit

Trehub S, Trainor L (1994) Les stratégies d'écoute chez le bébé: Origines du développement de la musique et de la parole. In Penser les sons, ed. S. MacAdams and E. Bigand. Paris: PUF

Wiesel TN, Hubel OH (1965) Comparison of the effects of unilateral and bilateral eye closure on cortical unit responses in kittens. J Neurophysiol 28: 1029–1040

Zeki S (1993) A vision of the brain. Oxford: Blackwell

Did Evolution Fix Human Values?

Camilo J. Cela-Conde[1]

This conference on "Neurobiology of Human Values," organised by the Fondation Ipsen, focuses on some very important aspects of the ever-controversial relationships between what we call "mind" and the brain. The title of its first session, "From aesthetics to ethics," expresses in itself the aim of reaching the core components of the human evaluative capacity.

Let me start with some assumptions that, due to the lack of space, I cannot explain in detail. I am sure thex will be easily accepted by readers interested in the neurobiology of values. First, moral and aesthetic judgements are functional states of the brain's activity. Nowadays, this is a very common way to conceive mental activity. Second, the existence and evolution of ethics – and aesthetics – might be better understood if we could identify the neural networks involved in moral and aesthetic judgements.

However, these assumptions pose some philosophical difficulties. Aesthetics and ethics embody an important part of human nature developed through phylogeny as a result of evolution by natural selection. On the other hand, the fields of ethics and aesthetics include ethical values and aesthetic preferences. What does that mean? Are these values also the result of evolution? Does human nature include the content of ethics and aesthetics, I mean, ethical and aesthetic values too?

To link genetic components and human values is a forbidden operation, given that it leads to the so-called naturalistic fallacy, formulated by David Hume. That is, if we deduce values from facts, we are making a logical mistake. Let me quote how Joshua Greene, who participates in this conference, expresses this advice: "There is a sharp and crucial distinction between the 'is' of science and the 'ought' of ethics" (Greene 2003). However, in the same paragraph, Greene points out that scientific statements could help us to re-evaluate our concepts of morality. To what extent are those statements helpful? And, particularly, how far should we re-evaluate our conceptions of morality to accept this kind of help?

In spite of philosophers' reluctance to accept values as a natural phenomenon, let us consider the following list of social regularities that are present in many non-human primates: food sharing, reciprocity of alliances, mutual assistance, retributive justice, reconciliation, consolation and conflict mediation.

---

[1] Laboratory of Human Systematics, University of the Balearic Islands, 7071 Palma de Mallorca, Spain; e-mail: cjcela@atlas-iap.es

Changeux et al.
Neurobiology of Human Values
© Springer-Verlag Berlin Heidelberg 2005

These behavioural traits could be considered, in Jessica Flack and Frans de Waal's (2000) opinion, as a sense of social regularity anticipating the human "moral sense."

Flack and de Waal's evolutionary sketch of the human capacities for moral judgment is not scientific nonsense lacking philosophical background. The Scottish Enlightenment understood that the "moral sense" was the element that, based on sympathy, leads human ethical choice. In his account of the evolution of co-operative behaviour, Darwin (1871) stated that any animal whatsoever, with well-defined social instincts – like parental and filial affections – "would inevitably acquire a moral sense or conscience, as soon as its intellectual powers had become as well, or nearly as well developed, as in man" (*Descent of Man*, p. 472). However, this is a hypothetical issue: no animal has ever reached the level of human mental faculties, language included. In fact, Darwin points out that, even if some animal could achieve a human-equivalent degree of development of its intellectual faculties, we could not conclude that it would also acquire exactly the same moral sense as ours (*Descent of Man*, p. 473). Therefore, human moral behaviour is a product of natural selection, but it is humankind's exclusive attribute.

Or is it?

The final step towards human values would have hypothetically been reached in our own species, by means of some cerebral modifications fixed through phylogenesis. As Terrence Deacon (1997) held, higher cognitive capacities would have evolved in our own species due to an enlargement of the prefrontal cortex. Thus, the cortical expansion responsible for developing values should have occurred after the cladistic episode that separated human and Neandertal lineages, that is to say, about 700,000 years ago. Actually, we have some items of evidence of high cognitive behaviour in the ancient burials of modern human beings. The burials are, in themselves, proof of the existence of ethical beliefs at that time, from the Upper Palaeolithic to the Neolithic. However, some decorative objects, such as necklaces and colour pigments, suggest the existence of an aesthetical sense too.

What had happened to the human brain by that time? Katherine Semendeferi and Hanna Damasio (2000) proved that no extra-allometric expansion exists in the frontal area of humans. In relative terms, we have the same parieto-occipital, temporal and frontal volumes corresponding to any ape. Only the overall large capacity of our brain makes a volumetric difference. However, the human brain is not just bigger than that of any other simian. Let us consider the cortical gyrification, measured by tracing an inner contour of the complete surface of the brain and by tracing an outer contour with tangential lines connecting the crests of gyral curvatures of the cortex. James Rilling and Thomas Insel (1999) studied the gyrification index of the primate cortex, showing that gyrification is allometric; the bigger the brain is, the more gyrificated the cortex is. However, an extra-allometric gyrification does exist, precisely in the prefrontal cortex of humans.

Could there be a relationship between the changes of the human frontal brain areas and the emergence of values? Since the proposal of the somatic marker hypothesis (Damasio 1994), a variety of articles confirmed the existence of a circuit linking the prefrontal and medial temporal lobes (Adolphs et al. 1998, Bechara et al. 2000; Moll et al. 2002; Keightley et al. 2003), that is, a network that is activated during moral judgments. As many of the participants at our conference will maintain, neuroimaging techniques can be used to localise some of these networks. However, what to say about aesthetics? Does the frontal cortex have any role in the human sense of aesthetics?

Our research team tried to answer this question by means of an experiment of localisation of aesthetic perception, carried out using magnetoencephalography (MEG). MEG is a type of brain imaging technique that measures the magnetic fields created by post-synaptic potentials. It has some important advantages: it is a non-invasive procedure, and it has very high temporal resolution and good spatial resolution combined with magnetic resonance (Cela-Conde et al. 2004).

In our experiment, eight female participants successively looked at a great number of pictures (320), deciding whether each picture was "beautiful" or not. Thus, participants themselves fixed the final aesthetic condition of every stimulus. High art pictures, both realistic and abstract, decorative pictures, and photographs of urban and natural landscapes – that is, highly varied material – were used as stimuli.

The results of our experiment show that the left prefrontal dorsolateral cortex (PDC) was greatly activated when participants perceived beautiful stimuli (either natural or artistic), reaching statistically significant differences (see Fig. 1). It is interesting to note that this differential activation took place at latencies ranging from 400 milliseconds to one second. We thus know that the frontal cortex is also activated in the perception of aesthetic pictures. However, we still have not studied the potential existence of invariants, that is, "universals," that could act as items of evidence of the evolutionary fixing of some aesthetic values. So, let's go back to the moral issue.

As Alan Sanfey and collaborators (2003) reported, the prefrontal dorsolateral cortex is activated in a task of evaluating fair and unfair exchanges, such as the so-called "Ultimatum game." Dominique de Quervain and collaborators (2004) also reported that the ventromedial prefrontal cortex and the medial orbitofrontal cortex are also activated when trying to integrate the benefits and costs of punishing selfish members of the group. On the grounds of both Sanfey's and de Quervain's experiments, it could be said that the value of fairness depends on the particular circuitry of certain neural networks in our brains. However, we are not the only primates with a sense of fairness. A similar criterion seems to be followed by the brown capuchin monkey, *Cebus apella*, which also return help in obtaining food (de Waal and Berger 2000) and reject an exchange of tokens for food in unequal conditions (Brosnan and de Waal 2003). Thus, capuchin monkeys are likely to have a sense of fairness, something that is considered as the main component of human justice by philosophers like John Rawls (1975).

What might be the evolutionary interpretation of that sense of fairness shared with other primates? Could this behaviour be considered a primitive trait fixed during primate phylogeny several million of years ago? To accept the homology, other primates, such as howling monkeys, baboons, vervet monkeys, gibbons, orangutans, gorillas and chimpanzees, should have the same-shared character of fairness (see Fig. 2). However, as de Waal reported in his book *Good Natured* (1996, p. 94), only capuchins, humans and chimpanzees seem to show this sense of fairness.

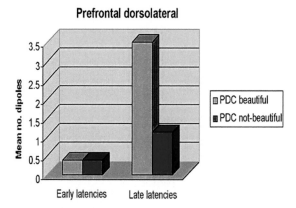

Fig. 1. Prefrontal dorsolateral cortex (PDC) activation in early latencies (100–400 ms) and late latencies (400–1000 ms) when participants see both natural and artistic pictures qualified as "beautiful" or "not-beautiful" by participants themselves. The results show an activation of PDC under the "beautiful" condition in late latencies, with statistically significant differences.

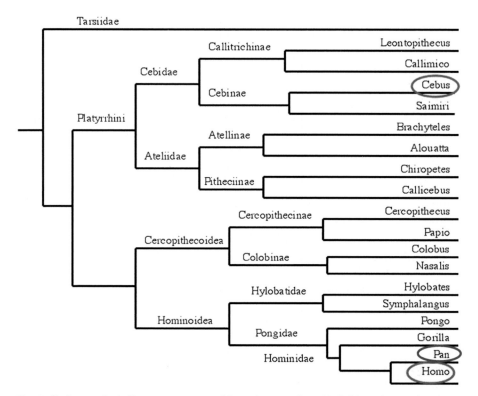

Fig. 2. Cladogram including some genera of the Primate order. Circled in red, taxa showing a "sense of fairness."

This evidence is too weak to justify the title of my contribution to this conference. As final conclusions, I think that we can agree on the existence of an evolutionary ground of human values. Neurobiology can also provide useful items of evidence about how human values work. However, we are quite far from being able to understand the evolution of the set brain/values. Much scientific work is still needed, in my opinion, to accurately explain the evolution of aesthetics and ethics.

References

Adolphs R, Tranel D, Damasio A (1998) The human amygdala in social judgement. Nature 393: 470–474.

Bechara A, Damasio H, Damasio A.R (2000) Emotion, decision making and the orbitofrontal cortex. Cereb Cortex 10: 295–307.

Brosnan S, De Waal F (2003) Monkeys reject unequal pay. Nature 425: 297–299.

Cela-Conde CJ, Marty G, Maestú F, Ortiz T, Munar E, Fernández A, Roca M, Rosselló J, Quesney F (2004) Activation of the prefrontal cortex in human aesthetic perception. Proc Natl Acad Sci USA 101: 6321–6325.

Damasio AR (1994) Descartes' error. Emotion, reason, and the human brain. New York, NY: G.P. Putnam's Sons.

Darwin C (1871) The descent of man, and selection in relation to sex. John Murray, London.

de Quervain DJ-F, Fischbacher U, Treyer V, Schellhammer M, Schnyder U, Buck A, Fehr E (2004) The neural basis of altruistic punishment. Science 305: 1254–1258.

de Waal F (1996) Good natured. The origins of right and wrong in humans and other animals. Cambridge, MA: Harvard University Press.

de Waal FBM, Berger ML (2000) Payment for labour in monkeys. Nature 404: 563.

Deacon T (1997) The symbolic species. New York; WW Norton & Co.

Flack JC, de Waal FBM (2000) Any animal whatever. J Consciousness Stud 7: 1–29.

Greene J (2003) From neural 'is' to moral 'ought': what are the moral implications of neuroscientific moral psychology. Neuroscience 4: 847–850.

Keightley ML, Winocur G, Graham SJ, Mayberg HS, Hevenor SJ, Grady CL (2003) An fMRI study investigating cognitive modulation of brain regions associated with emotional processing of visual stimuli. Neuropsychologia 41: 585–596.

Moll J, Oliveira-Souza R, Bramati I, Grafman J (2002) Functional networks in emotional moral and nonmoral social judgements. NeuroImage 16: 696–703.

Rawls J (1975) A theory of justice. Cambridge, MA: Harvard University Press.

Rilling JK, Insel TR (1999) The primate neocortex in comparative perspective using magnetic resonance imaging. J Human Evol 37: 191–223.

Sanfey AG, Rilling JK, Aronson JA, Nystrom LE, Cohen JD (2003) The neural basis of economic decision-making in the ultimatum game. Science 300: 1755–1758.

Semendeferi K, Damasio H (2000) The brain and its main anatomical subdivisions in living hominoids using magnetic resonance imaging. J Human Evol 38: 317–332.

Homo homini lupus? Morality, the Social Instincts, and our Fellow Primates

Frans B. M. de Waal[1]

> Why should our nastiness be the baggage of an apish past
> and our kindness uniquely human?
> Why should we not seek continuity with other animals
> for our 'noble' traits as well?"
>
> Stephen Jay Gould (1980, p. 261)

The question whether we are naturally good or bad is a perennial one. The past quarter of a century has seen biologists weigh in with a thoroughly pessimistic view according to which we are not naturally moral. This is a curious position, because it presents us as going against our own nature when striving for a moral life. Accordingly, we are moral only thanks to culture and religion. I will call this "Veneer Theory," as it postulates morality as a thin layer disguising the less noble tendencies seen as the true core of human nature (de Waal 2005). Veneer Theory radically breaks with Charles Darwin's position that morality is an outgrowth of the social instincts, hence continuous with the sociality of other animals (Darwin 1871).

Veneer Theory suffers from several unanswered questions. First, why would humans, and humans alone, have broken with their own biology? Second, how is such a feat even possible? And third, where are the empirical data? The theory predicts that morality resides in recently evolved parts of our enlarged brain and leads to behavior substantially at odds with that of animals. Both are verifiably wrong. For the latest human evidence, I refer to Haidt (2001), Greene and Haidt (2002) and Greene (2005). This chapter discusses the animal evidence. Animals may not be moral beings, but they do show signs of empathy, reciprocity, and a sense of fairness. I will review evidence for the continuity between humans and other primates expected by Darwin, and suggest that the building blocks of morality are older than our species.

Veneer Theory

In 1893, Thomas Henry Huxley publicly tried to reconcile his dim view of a nasty natural world with the kindness occasionally encountered in human society. Hux-

[1] Living Links, Yerkes Primate Center, Emory University, 954 N. Gatewood Road, Atlanta GA 30322; e-mail: dewaal@emory.edu

Changeux et al.
Neurobiology of Human Values
© Springer-Verlag Berlin Heidelberg 2005

ley realized that the laws of the physical world are unalterable. He felt, however, that their impact on human existence could be softened and modified provided people kept nature under control. He compared us with a gardener who has a hard time keeping weeds out of his garden. Thus, he proposed ethics as a human cultural victory over the evolutionary process in the same way as the gardener conquers the weeds in his garden (Huxley 1894).

This position deliberately curbed the explanatory power of evolution. Since many consider morality the essence of our species, Huxley was in effect saying that what makes us human could not be handled within an evolutionary framework. It was an inexplicable retreat by someone who had gained a reputation as "Darwin's Bulldog," owing to his fierce advocacy of evolutionary theory. His solution was quintessentially Hobbesian in that it stated that people are fit for society only by education, not their nature (Hobbes 1651).

If we are indeed Hobbesian competitors who don't care one bit about the feelings of others (*Homo homini lupus,* or Man is wolf to man), how did we decide to transform ourselves into model citizens? Can people for generations maintain behavior that is out of character, like a shoal of piranhas that decide to become vegetarians? How deep does such a change go?

Darwin saw morality in a totally different light. As Huxley's biographer, Desmond (1994, p. 599), put it: "Huxley was forcing his ethical Ark against the Darwinian current which had brought him so far." Two decades earlier, in *The Descent of Man,* Darwin (1871) had unequivocally stressed continuity between human nature and morality. The reason for Huxley's departure has been sought in his suffering from the cruel hand of nature, which had taken the life of his daughter, as well as his need to make the ruthlessness of the Darwinian cosmos palatable to the general public. He could do so only, he felt, by dislodging human ethics, declaring it a cultural innovation (Desmond 1994).

Huxley's dualism was to get a respectability boost from Sigmund Freud's writings, which thrived on contrasts between the conscious and subconscious, the ego and super-ego, Love and Death, and so on. As with Huxley's gardener and garden, Freud was not just dividing the world in symmetrical halves: he saw struggle everywhere. He let civilization arise out of a renunciation of instinct, the gaining of control over the forces of nature, and the building of a cultural super-ego (Freud 1930).

That this remains a theme today is obvious from the statements by outspoken Huxleyans, who are still found among biologists. Declaring ethics a radical break with biology, George Williams has written extensively about the wretchedness of nature, culminating in his claim that human morality is an inexplicable accident of the evolutionary process: "I account for morality as an accidental capability produced, in its boundless stupidity, by a biological process that is normally opposed to the expression of such a capability" (Williams 1988, p. 438).

Richard Dawkins (1996) suspects that we are nicer than is good for our "selfish genes," and has explicitly endorsed Huxley: "What I am saying, along with many other people, among them T. H. Huxley, is that in our political and social life we are entitled to throw out Darwinism, to say we don't want to live in a Darwinian world" (Roes 1997, p. 3).

Like Huxley, these authors generally believe that human behavior is an evolutionary product, except when it is unselfish. And like Freud, they propose dichotomies: we are part nature, part culture, rather than a well-integrated whole. The same position has been echoed by popularizers such as Robert Wright (1994), who, in *The Moral Animal,* went so far as to claim that virtue is absent from people's hearts and souls. He bluntly stated that our species is potentially but not naturally moral. But what if people occasionally experience in themselves and others a degree of sympathy, goodness, and generosity? Wright's answer is that the "moral animal" is a fake:

"… the pretense of selflessness is about as much part of human nature as is its frequent absence. We dress ourselves up in tony moral language, denying base motives and stressing our at least minimal consideration for the greater good; and we fiercely and self-righteously decry selfishness in others." (Wright 1994, p. 344).

This statement recalls the famous synopsis of Veneer Theory by Ghiselin (1974, p. 247): "Scratch an 'altruist,' and watch a 'hypocrite' bleed." To explain how we manage to live with ourselves despite this travesty, theorists have called upon self-deception (e.g. Badcock 1986). The problem is, of course, that self-deception is a most cognitively demanding process compared with the sentiments proposed by Hume (1739), Darwin (1871), Westermarck (1906, 1908), and me (de Waal 1996a; Flack and de Waal 2000), which make morality flow naturally from inborn social tendencies. To illustrate these tendencies, I will review evidence for conflict resolution, empathy, reciprocity, and an aversion to unfairness in nonhuman primates.

Conflict Resolution

Reconciliation

In the summer of 2002, various national European behavioral biology and ethology societies came together for a conference on animal conflict resolution. This field started out with simple descriptive work but is now rapidly moving towards a theoretical framework supported by observational as well as experimental data (for reviews, see de Waal 2000; Aureli and de Waal 2000; Aureli et al. 2002).

Reconciliation was first reported by de Waal and van Roosmalen (1979). A typical example concerns two male chimpanzees who have been chasing each other, barking and screaming, and afterwards rest in a tree (Fig. 1). Ten minutes later, one male holds out his hand, begging the other for an embrace. Seconds later, they hug and kiss, and climb down to the ground together to groom each other. Termed a "reconciliation," this process is defined as a friendly contact not long after a conflict between two parties. A kiss is the most typical way for chimpanzees to reconcile. Other animals have different styles. Bonobos do it with sex, and stumptail macaques wait till the subordinate presents, then hold its hips in a so-called hold-bottom ritual. Each species has its own way, yet the basic principle remains the same, which is that former opponents reunite following a fight.

Fig. 1. Reconciliation 10 minutes after a protracted, noisy conflict between two adult males at the Arnhem Zoo. The challenged male (left) had fled into the tree, but 10 minutes later his opponent stretched out a hand. Within seconds, the two males had a physical reunion. Photograph by the author.

Primatology has always shown interest in social relationships, so the idea of relationship repair, implied by the reconciliation label, was quickly taken seriously. We now know that about 30 different primate species reconcile after fights, and recent studies show that reconciliation is not at all limited to the primates. There is evidence for this mechanism in hyenas, dolphins, and domestic goats (Schino 2000). Reconciliation seems a basic process found, or to be found, in a host of social species. The reason for it being so widespread is that it restores relationships that have been disturbed by aggression but are nonetheless essential for survival. Since many animals establish cooperative relationships within which conflict occasionally arises, many need mechanisms of repair.

A standard research method is the post-conflict/matched control (PC/MC) method (de Waal and Yoshihara 1983). Observations start with a spontaneous aggressive encounter after which the combatants are followed for a fixed period of time, say 10 minutes, to see what subsequently happens between them. This is the PC or post-conflict observation. Figure 2, which concerns stumptail macaques, shows that approximately 60% of the pairs of opponents come together after a fight. This finding is compared with control observations that tell us how these monkeys normally act without a preceding fight. Since control observations are done on a different observation day but matched to the PC observation for the time of the day and the individuals involved, they are called MCs or matched controls.

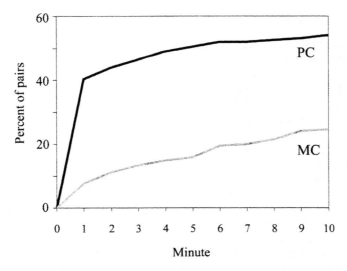

Fig. 2. Primates show a dramatic increase in body contact between former opponents during post-onflict (PC) as compared to matched-control (MC) observations. The graph provides the cumulative percentage of opponent-pairs establishing friendly contact during a 10-min time window following 670 spontaneous aggressive incidents in a zoo group of stumptail macaques. Based on de Waal and Ren (1988).

Notice that there is far more contact after fights than in the control observations, which is exactly the opposite picture from that presented by the textbooks I read as a graduate student. In those days, Lorenz's (1963) *On Aggression* was influential. The popular idea was that aggression is a dispersive behavior, which serves to space out individuals. This idea was developed on territorial species, which were the first studied. With social animals, however, things are quite different. In primates, we actually see the opposite: aggression literally brings individuals together.

If the same observations and analyses are conducted on human children, as a co-worker and I did at a preschool near our university, one finds the familiar PC/MC pattern (Verbeek and de Waal 2001). An extensive review of recent child studies by Verbeek et al. (2000) confirms that the data look essentially the same for children, chimpanzees, monkeys, and goats. After fights, individuals come together more than normally, often with intense contact patterns, doing so especially with partners whom they need for one reason or another. The latter is known as the Valuable Relationship Hypothesis, which can be formulated thus: "Reconciliation will occur especially between individuals who stand much to lose if their relationship deteriorates." This hypothesis is well-supported by observational studies as well as by an elegant experiment on monkeys, which manipulated relationship value by promoting cooperation (Cords and Thurnheer 1993).

The Relational Model

Peace is not sought for peace's sake but in order to preserve mutual interests. The same principle is known in human affairs. For example, the idea underlying the European Community was that nations with a recent history of mutual warfare may show an increased tendency to keep the peace if they are made mutually dependent on each other. Europeans have worked on increasing relationship value since World War II, recently culminating in the adoption of a common currency.

I have formalized the above ideas in the Relational Model, which places conflict in a social context. Instead of aggression being an instinct or an automatic response triggered by frustration, it is one of several options for the resolution of conflicts of interest. Other options are avoidance of the adversary (common in hierarchical and territorial species) and the sharing of resources (common in tolerant species). Weighing the costs and benefits of each option, conflict may escalate to the point of aggression, after which there still is the option of undoing its damage by means of reconciliation, which option will be most favored by parties with overlapping interests (de Waal 1996b, 2000).

Empathy

Emotional Linkage in Animals

When Zahn-Waxler visited homes to find out how children respond to family members who were instructed to feign sadness (sobbing), pain (crying), or distress (choking), she discovered that children a little over one year of age already comfort others. This is a milestone in their development: an aversive experience in another person draws out a concerned response. An unplanned side-bar to this classical study, however, was that some household pets appeared as worried as the children by the "distress" of a family member. They hovered over them or put their heads in their laps (Zahn-Waxler et al. 1984).

Intersubjectvitity has many aspects apart from emotional linkage, such as an appraisal of the other's situation, experience-based predictions about the other's behavior, extraction of information from the other that is valuable to the self, and an understanding of the other's knowledge and intentions. When the emotional state of one individual induces a matching or related state in another, we speak of emotional contagion (Hatfield et al. 1993). With increasing differentiation between self and other, and an increasing appreciation of the precise circumstances underlying the emotional states of others, emotional contagion develops into empathy. Empathy encompasses – and could not possibly exist without – emotional contagion, yet empathy goes beyond it in that it places filters between the other's state and the subject's own state and adds a cognitive layer. In empathy, the subject does *not* confuse its own internal state with the other's. These various levels of empathy, including personal distress and sympathetic concern, are defined and discussed in detail by Eisenberg (2000).

Empathy is a social phenomenon with great adaptive significance for animals in groups. The fact that most modern textbooks on animal cognition do not index empathy or sympathy does not mean that these capacities are not essential; it only means that they have been overlooked by a science traditionally focused on individual rather than inter-individual capacities. Inasmuch as the survival of many animals depends on concerted action, mutual aid, and information transfer, selection must have favored proximate mechanisms to evaluate the emotional states of others and quickly respond to them in adaptive ways. Even though the human empathy literature often emphasizes the cognitive side of this ability, Hoffman (1981, p. 79) rightly noted early on that "humans must be equipped biologically to function effectively in many social situations without undue reliance on cognitive processes."

Empathy allows us to relate to the emotional states of others. This ability is critical for the regulation of social interactions, such as coordinated activity, cooperation towards a common goal, social bonding, and care of others. It would be strange indeed if a survival mechanism that arises so early in life in all members of our species would be without animal parallels.

Early Experiments

An interesting older literature by experimental psychologists addresses empathy (reviewed by Preston and de Waal 2002a,b). In a paper provocatively entitled "Emotional reactions of rats to the pain of others," Church (1959) established that rats that had learned to press a lever to obtain food would stop doing so if their response was paired with the delivery of an electric shock to a neighboring rat. Even though this inhibition habituated rapidly, it suggested something aversive about the pain reactions of others. Perhaps such reactions arouse negative emotions in those who see and hear them.

In the same year as Church's rat study, Miller et al. (1959) published the first of a series of pioneering papers on the transmission of affect in rhesus macaques. They found that monkeys react with avoidance to pictures of conspecifics in a fearful pose, and that this reaction is stronger than that towards a negatively conditioned stimulus. This discovery was astonishing, suggesting that seeing the fear of a two-dimensional, soundless representation of another monkey is more disturbing than the anticipation of an actual electric shock.

Perhaps the most compelling evidence for the strength of the empathic reaction in monkeys came from Wetchkin et al. (1964) and Masserman et al. (1964). They found that rhesus monkeys refuse to pull a chain that delivers food to themselves if doing so shocks a companion. One monkey stopped pulling for five days, and another one for twelve days, after witnessing shock-delivery to a companion. These monkeys were literally starving themselves to avoid inflicting pain upon another, and they maintained this response to a far greater degree than has been reported for rats.

Evidence of Primate Empathy

Qualitative accounts of great apes support the view that these animals show strong emotional reactions to others in pain or need. Yerkes (1925, p. 246) reported how his bonobo, Prince Chim, was so extraordinarily concerned and protective towards his sickly chimpanzee companion, Panzee, that the scientific establishment might not accept his claims: "If I were to tell of his altruistic and obviously sympathetic behavior towards Panzee I should be suspected of idealizing an ape" (Fig. 3). Ladygina-Kohts (1935, p. 121) noticed similar empathic tendencies in her young chimpanzee, Joni, whom she raised at the beginning of the previous century, in Moscow. Kohts, who analyzed Joni's behavior in the minutest detail, discovered that the only way to get him off the roof of her house after an escape

Fig. 3. Robert Yerkes with two apes in his lap. This photo was taken in 1923, before the discovery of the bonobo as a distinct species. We know now that the ape on the right, named Prince Chim, was a bonobo. Prince Chim was described by Yerkes as gentler and more empathic than any other ape he knew. In debates about human evolution, the bonobo is largely ignored, however. (Photograph by Lee Russell, courtesy of the Yerkes National Primate Research Center).

(much better than any reward or threat of punishment) was by appealing to his sympathy:

"If I pretend to be crying, close my eyes and weep, Joni immediately stops his plays or any other activities, quickly runs over to me, all excited and shagged, from the most remote places in the house, such as the roof or the ceiling of his cage, from where I could not drive him down despite my persistent calls and entreaties. He hastily runs around me, as if looking for the offender; looking at my face, he tenderly takes my chin in his palm, lightly touches my face with his finger, as though trying to understand what is happening, and turns around, clenching his toes into firm fists."

Similar reports are discussed by de Waal (1996a, 1997a), who suggests that, apart from emotional connectedness, apes have an appreciation of the other's situation and a degree of perspective-taking. Apes show the same empathic capacity that was so enduringly described by Smith (1759, p. 10) as "changing places in fancy with the sufferer."

O'Connell (1995) conducted a content analysis of thousands of qualitative reports, counting the frequency of three types of empathy, from emotional contagion to more cognitive forms, including an appreciation of the other's situation. Understanding the emotional state of another was particularly common in the chimpanzee, with most outcomes resulting in the subject comforting the object of distress. Monkey displays of empathy were far more restricted but did include the adoption of orphans and reactions to illness, handicaps, and wounding.

This difference between monkey and ape empathy is also evident from systematic studies of behavior known as "consolation," first documented by de Waal and van Roosmalen (1979). Consolation is defined as friendly, reassuring contact directed by an uninvolved bystander at one of the combatants in a previous aggressive incident. For example, a third party goes over to the loser of a fight and gently puts an arm around his shoulders (Fig. 4). Consolation is not to be confused with

Fig. 4. Consolation among chimpanzees: a juvenile puts an arm around a screaming adult male who has just been defeated in a fight with a rival. (Photograph by the author).

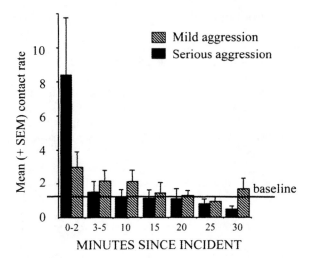

Fig. 5. The rate with which third parties contact victims of aggression in chimpanzees, comparing recipients of serious and mild aggression. Especially in the first two minutes following the incident, recipients of serious aggression receive more contacts than baseline. After de Waal and Aureli (1996).

reconciliation, which seems selfishly motivated (see above). The advantages of consolation for the actor remain unclear. The actor could probably walk away from the scene without negative consequences.

Chimpanzee consolation is well quantified. De Waal and van Roosmalen (1979) based their conclusions on an analysis of hundreds of post-conflict observations, and a replication by de Waal and Aureli (1996) included an even larger sample in which the authors sought to test two relatively simple predictions. If third-party contacts indeed serve to alleviate the distress of conflict participants, these contacts should be directed more at recipients of aggression than at aggressors, and more at recipients of intense rather than mild aggression. Comparing third-party contact rates with baseline levels, we found support for both predictions (Fig. 5).

Consolation has thus far been demonstrated only in great apes. When de Waal and Aureli (1996) set out to apply exactly the same observation protocols as used on chimpanzees to detect consolation in macaques, they failed to find any (see also Watts et al. 2000). This finding came as a surprise, because reconciliation studies, which employ essentially the same design, have shown reconciliation in species after species. Why, then, would consolation be restricted to apes?

Targeted helping in response to specific, sometimes novel situations may require a distinction between self and other that allows the other's situation to be divorced from one's own while maintaining the emotional link that motivates behavior. Possibly, one cannot achieve cognitive empathy without a high degree of self awareness. In other words, to understand that the source of a vicarious affective state is not oneself but the other, and to understand why the other's state arose (e.g., the specific cause of the other's distress), one needs a distinction between self and other. Based on these assumptions, Gallup (1982) was the first to speculate about a possible connection between cognitive empathy and mirror self-recognition (MSR). This view is supported both developmentally, by a correlation between the emergence of MSR in children and their helping tendencies

(Bischof-Köhler 1988; Zahn-Waxler et al. 1992), and phylogenetically, by the presence of complex helping and consolation in hominoids (i.e., humans and apes) but not monkeys. Hominoids are also the only primates with MSR.

I have argued before that, apart from consolation behavior, targeted helping reflects cognitive empathy. Targeted helping is defined as altruistic behavior tailored to the specific needs of the other even in novel situations, such as the highly publicized case of Binti-Jua, a gorilla female who rescued a human boy at the Brookfield Zoo, in Chicago (de Waal 1996a, 1999). Targeted helping is common in the great apes but is also striking in dolphins (Caldwell and Caldwell, 1966). The recent discovery of MSR in dolphins (Reiss and Marino, 2001) thus fits the proposed connection between increased self-awareness, on the one hand, and cognitive empathy, on the other.

Russian Doll model

The literature includes accounts of empathy as a cognitive affair, even to the point that apes, let alone other animals, probably lack it (Povinelli 1998; Hauser 2000). This "top-down" view equates empathy with mental state attribution and theory-of-mind (ToM). The opposite position has recently been defended, however, in relation to autistic children. In contrast to earlier assumptions that autism reflects a deficit in ToM (Baron-Cohen 2000), autism is noticeable well before the age of four, at which time ToM typically emerges. Williams et al. (2001) argue that the main deficit of autism concerns the socio-affective level, which then negatively impacts more sophisticated down-stream forms of inter-personal perception, such as ToM (see also Baron-Cohen 2004).

Preston and de Waal (2002a) propose that at the core of the empathic capacity is a relatively simple mechanism that provides an observer (the "subject") with access to the subjective state of another (the "object") through the subject's own neural and bodily representations. When the subject attends to the object's state, the subject's neural representations of similar states are *automatically* activated. The more similar the subject and object, the more the object will activate matching peripheral motor and autonomic responses in the subject (e.g., changes in heart rate, skin conductance, facial expression, body posture). This activation allows the subject to understand that the object also has an extended consciousness, including thoughts, feelings, and needs, which in turn fosters sympathy, compassion, and helping. Preston and de Waal (2002b) and de Waal (2003) propose evolutionary continuity between humans and other mammals in this regard. Their Perception-Action Model (PAM) fits Damasio's (1994) somatic marker hypothesis of emotions as well as recent evidence for a link at the cellular level between perception and action (e.g,. "mirror neurons;" di Pelligrino et al. 1992).

The idea that perception and action share common representations is anything but new: it goes as far back as the first treatise on "Einfühlung," the German concept later translated into English as "empathy" (Wispé 1991). When Lipps (1903) introduced Einfühlung, which literally means "feeling into," he speculated about "innere Nachahmung" (inner mimicry) of another's feelings along the same lines as proposed by the PAM. Accordingly, empathy is often an automatic, insuppressible process, as demonstrated by electromyographic studies of invisible muscle contractions in people's faces in response to pictures of human facial expressions (Dimberg et al. 2000). Accounts of empathy as a higher cognitive process neglect these "gut level" reactions, which are far too rapid to be under voluntary control.

Perception-action mechanisms are well known for motor perception (Prinz and Hommel, 2002), causing researchers to assume that similar processes underlie emotion perception (Gallese 2001; Wolpert et al. 2001). Data suggest that observing and experiencing an emotion involve similar physiological substrates (Adolphs et al. 1997, 2000), and that affect communication creates similar physiological activity in subject and object (Dimberg 1982, 1990; Levenson and Reuf, 1992). Recent investigations of neural mechanisms of empathy (Carr et al. 2003; Preston et al, 2002; Wicker et al., 2003; Singer et al. 2004; de Gelder et al, 2004) lend support to, or are at least consistent with, the PAM.

This "bottom-up" view was depicted as a Russian doll by de Waal (2003). Accordingly, empathy covers all forms of one individual's emotional state affecting another's, with simple mechanisms at its core and more complex mechanisms, cognitive filters, and perspective-taking abilities built on top. Autism may be reflected in deficient outer layers of the Russian doll, but such deficiencies most likely go back to deficient inner layers.

This is not to say that higher cognitive levels of empathy are irrelevant, but they are built on top of this firm, hard-wired basis without which we would be at a loss about what moves others. Thus, at the core of the Russian doll we find emotional contagion around which cognitive empathy and attribution are constructed. Cognitive empathy implies appraisal of another's predicament or situation (cf. de Waal,1996a). The subject not only responds to the signals emitted by the object but seeks to understand the reasons for these signals, looking for clues in the other's behavior and circumstances. Cognitive empathy makes it possible to furnish targeted helping that takes the needs of the other into account. These responses go well beyond emotional contagion, yet they would be hard to explain without an emotional motivational component.

Whereas monkeys (and many other social mammals) clearly seem to possess emotional contagion and some forms of targeted helping, the latter phenomenon is not nearly as robust as in the great apes. For example, at Jigokudani monkey park, in Japan, first-time mother macaques are kept out of the hot water springs by park wardens because of their experience that these females will accidentally drown their infants. They fail to pay attention to them when submerging themselves in the ponds (de Waal 1996b). This behavior is something they apparently learn to do over time, showing that they do not automatically take their offspring's perspective. Ape mothers, in contrast, respond immediately and appropriately to the specific needs of their offspring and are very careful to keep them away from water (personal observations).

In conclusion, empathy is not an all-or-nothing phenomenon: it covers a wide range of emotional linkage patterns, from the very simple and automatic to the very sophisticated. It seems logical to first try to understand the more basic forms, which are widespread indeed, before addressing the interesting variations that cognitive evolution has constructed on top of this foundation.

Reciprocity

Chimpanzees and capuchin monkeys – the two species I work with the most – are special, because they are among the very few primates that share food outside the mother-offspring context. The capuchin is a small, easy animal to work with, as opposed to the chimpanzee, which is many times stronger than we are. But chimpanzees, too, are interested in each others' food and will share on occasion – sometimes even hand over a piece of food to another. Most sharing, however, is passive, where one individual will reach for food owned by another, who will let go (de Waal 1997b). But even such passive sharing is special compared to most animals, in which such a situation might result in a fight or assertion by the dominant, without sharing.

One series of experiments concerned the idea that monkeys cooperate on the basis of mental record-keeping of favors. We set up a situation to study tit-for-tat: I do something for you and, a while later, you do something for me. Inspired by a classic 1930s study at the Yerkes Primate Center and the theories of Trivers (1971), we presented a pair of capuchin monkeys with a tray with two pull bars attached to it (de Waal and Berger 2000). Both monkeys sat in a test chamber with mesh between them, so that they could see each other and share food through the mesh. The tray was counterweighed such that a single monkey couldn't pull it: they needed to work together. Only one side was baited, meaning that only one of the two monkeys would obtain a food reward.

After successful pulls we measured how much food the possessor shared with its helper. Possessors could easily keep the food by sitting in the corner and eating alone, but didn't do so. We found that food sharing after cooperative efforts was higher than after solo efforts. That is, the possessor of food shared more often with the monkey on the other side of the mesh if this partner had played a role in securing the food than if the possessor had acquired the food on its own. Capuchins thus seem to reward helpers for their efforts, which is of course also a way of keeping assistants motivated.

Fairness

The above relates to the distribution of pay-offs. How skewed can it be before cooperation disappears? According to a recent theory, the well-known human aversion to inequity relates to the need to maintain cooperation (Fehr and Schmidt 1999). Similarly, cooperative nonhuman species seem guided by a set of expectations about pay-off distribution. De Waal (1996a, p. 95) proposed a *sense of* "social regularity," defined as "a set of expectations about the way in which oneself (or others) should be treated and how resources should be divided. Whenever reality deviates from these expectations to one's (or the other's) disadvantage, a negative reaction ensues, most commonly protest by subordinate individuals and punishment by dominant individuals."

Note that the expectations have not been specified: they are species-typical. To explore expectations held by capuchin monkeys, we made use of their ability to judge and respond to value. We knew from previous studies that capuchins easily learn to assign value to tokens. Furthermore, they can use these assigned values to complete a simple barter. This ability made possible a test to elucidate inequity aversion by measuring the reactions of subjects to a partner receiving a superior reward for the same tokens.

We paired each monkey with a group mate and watched reactions if their partners got a better reward for doing the same bartering task. This task consisted of an exchange in which the experimenter gave the subject a token that could immediately be returned for a reward. Each session consisted of 25 exchanges by each individual, and subjects always saw their partner's exchange immediately before their own. Food rewards varied from low-value rewards (a cucumber piece), which they are usually happy to work for, to high-value rewards (a grape), which were preferred by all individuals tested. All subjects were subjected to 1) an equity test, in which subject and partner did the same work for the same low-value food, 2) an inequity test, in which the partner received a superior reward (grape) for the same effort, 3) an effort control test, designed to elucidate the role of effort, in which the partner received the higher-value grape for free, and 4) a food control test, designed to elucidate the effect of the presence of the reward on subject behavior, in which grapes were visible but not given to another capuchin.

Individuals who received lower-value rewards showed both passive negative reactions (i.e., refusal to exchange the token, ignoring the reward) and active negative reactions (i.e., throwing out the token or reward). Compared to tests in which both received identical rewards, the capuchins were far less willing to complete the exchange or accept the reward if their partner received a better deal (Brosnan and de Waal 2003). Capuchins refused to participate even more frequently if their partner did not have to work (exchange) to get the better reward but was handed it for "free."

Of course, there is always the possibility that subjects were reacting to the mere presence of the higher-value food, and that what the partner received (free or not) did not affect their reaction. However, in the food control test, in which the higher-value reward was visible but not given to another monkey, the reaction to the presence of this high-valued food decreased significantly over the course of testing, which is a change in the opposite direction from that seen when the high-value reward went to an actual partner. Clearly, our subjects discriminate between higher-value food being consumed by a conspecific and such food being merely visible, intensifying their rejections only to the former (Brosnan and de Waal 2004).

Capuchin monkeys thus seem to measure reward in relative terms, comparing their own rewards with those available and their own efforts with those of others. Although our data cannot elucidate the precise motivations underlying these responses, one possibility is that monkeys, like humans, are guided by social emotions. These emotions, known as "passions" by economists, guide human reactions to the efforts, gains, losses, and attitudes of others (Hirschleifer 1987; Frank 1988; Sanfey et al, 2003). As opposed to primates marked by despotic hierarchies, tolerant species with well-developed food sharing and cooperation, such as capuchin monkeys, may hold emotionally charged expectations about reward distribution and social exchange that lead them to dislike inequity.

Conclusion

We have contrasted here two separate schools of thought on human goodness. One sees people as essentially evil and selfish and explains morality as a cultural overlay ungrounded in human nature or evolutionary theory. This dualistic school of thought, personified by T. H. Huxley, is still very much with us.

The second school, going back to David Hume and Charles Darwin (and ultimately Aristotle; Arnhart 1998), sees the moral sense naturally arising in our species. Apart from our obvious competitive tendencies, we are social to the core. The human species is what zoologists call "obligatorily social", that is, its survival is closely tied to group life and cooperation. Obviously, the question of how we came to be this way can be answered only if we broaden the evolutionary horizon beyond the dog-eat-dog theories that have dominated science writing about human biology over the past three decades.

The child is not going against its own nature by developing a caring, moral attitude any more than civil society is an out-of-control garden subdued by the sweating gardener of Huxley's imagination. Moral attitudes have been with us from the start, and the gardener is, as John Dewey aptly put it, an organic grower. The successful gardener creates conditions and introduces plants that may not be normal for this particular plot of land "but fall within the wont and use of nature as a whole" (Dewey 1898, pp. 109–110).

Two field of research can throw new light on the age-old problem of the origins of human morality. One is modern neuroscience, which can reveal how moral decision-making relates at a neural level to ancient social emotions. The second is research into animal behavior, with attention to the emotions. That is, instead of studying prosocial tendencies only in terms of their costs and benefits, as is commonly done, they should also be studied from a psychological perspective to know which stimulus situations induce these tendencies and what kind of cognition animals put into their execution.

References

Adolphs R, Cahill L, Schul R, Babinsky R (1997) Impaired declarative memory for emotional material following bilateral amygdala damage in humans. Learning Memory 4: 291–300.

Adolphs R, Damasio H, Tranel D, Cooper G, Damasio AR (2000) A role for somatosensory cortices in the visual recognition of emotion as revealed by three-dimensional lesion mapping. J Neurosci 20: 2683–2690.

Arnhart L (1998) Darwinian natural right: the biological ethics of human nature. Albany, NY: SUNY Press.

Aureli F, de Waal FBM (2000) Natural conflict resolution. Berkeley, CA: University of California Press.

Aureli F, Cords M, van Schaik CP (2002) Conflict resolution following aggression in gregarious animals: A predictive framework. Animal Behav 64: 325–343.

Badcock CR (1986) The problem of altruism: Freudian-Darwinian solutions. Oxford: Blackwell.

Baron-Cohen S (2000) Theory of mind and autism: A fifteen year review. In: Baron-Cohen S, Tager-Flusberg H, Cohen DJ (eds) Understanding other minds.. Oxford: Oxford University Press, pp. 3–20

Baron-Cohen S (2004). Sex differences in social development: Lessons from autism. In: Leavitt LA, Hall DMB (eds) Social and moral development: emerging evidence on the toddler years. Johnson & Johnson Pediatric Institute, pp. 125–141

Bischof-Köhler D (1988) Über den Zusammenhang von Empathie und der Fähigkeit sich im Spiegel zu erkennen. Schweiz Zeits Psycholog 47: 147–159.

Brosnan SF, de Waal FBM (2003) Monkeys reject unequal pay. Nature 425: 297–299.

Brosnan SF, de Waal FBM (2004) Reply to Henrich & Wynne. Nature 428: 140.

Caldwell MC, Caldwell DK (1966) Epimeletic (care-giving) behavior in Cetacea. In: Norris KS (ed) Whales, dolphins, and porpoises. Berkeley, CA : University of California Press, pp. 755–789.

Carr L, Iacoboni M, Dubeau MC, Mazziotta JC, Lenzi GL (2003) Neural mechanisms of empathy in humans: A relay from neural systems for imitation to limbic areas. Proc Natl Acad Sci USA 100: 5497–5502.

Church RM (1959) Emotional reactions of rats to the pain of others. J Comp Physiol Psychol 52: 132–134.

Cords M, Thurnheer S (1993) Reconciliation with valuable partners by long-tailed macaques. Ethology 93: 315–325.

Damasio A (1994) Descartes' error: emotion, reason, and the human Brain. New York: Putnam.

Darwin C (1981 [1871]) The descent of man, and selection in relation to sex. Princeton, NJ: Princeton University Press.

Dawkins R (1996) No title. Times Literary Supplement. November 29: 13.

de Gelder B, Snyder J, Greve D, Gerard G, Hadjikhani N (2004) Fear fosters flight: A mechanism for fear contagion when perceiving emotion expressed by a whole body. Proc Natl Acad Sci USA 101: 16701–16706.

de Waal FBM (1996a) Good natured: the origins of right and wrong in humans and other animals. Cambridge, MA: Harvard University Press.

de Waal FBM (1996b) Conflict as negotiation. In: McGraw WC, Marchant LF, Nishida T (eds) Great Ape Societies. Cambridge : Cambridge University Press, Cambridge, pp 159–172.

de Waal FBM (1997a) Bonobo: the forgotten ape. Berkeley, CA: University of California Press.

de Waal FBM (1997b) The chimpanzee's service economy: Food for grooming. Evol Human Behav 18: 375–386.

de Waal FBM (1999) Anthropomorphism and anthropodenial: consistency in our thinking about humans and other animals. Philosoph Topics 27: 255–280.

de Waal FBM (2000. Primates – A natural heritage of conflict resolution. Science 289: 586–590.

de Waal FBM (2003). On the possibility of animal empathy. In: Manstead T, Frijda N, Fischer A (eds) Feelings and emotions: the Amsterdam Symposium. Cambridge: Cambridge University Press pp 379–399.

de Waal F.BM (2005). Morality and the social instincts: continuity with the other primates. In: Peterson GB (ed) Tanner Lectures on Human Values. University of Utah Press.

de Waal FBM, Aureli F (1996) Consolation, reconciliation, and a possible cognitive difference between macaque and chimpanzee. In: Russon AE, Bard KA Parker ST (eds) Reaching into thought: the minds of the great apes. Cambridge: Cambridge University Press, pp. 80–110.

de Waal FBM, Berger ML (2000) Payment for labour in monkeys. Nature 404: 563.

de Waal FBM, Ren R (1988) Comparison of the reconciliation behavior of stumptail and rhesus macaques. Ethology 78: 129–142.

de Waal FBM, van Roosmalen A (1979) Reconciliation and consolation among chimpanzees. Behav Ecol Sociobiol 5: 55–66.

de Waal FBM, Yoshihara D (1983) Reconciliation and re-directed affection in rhesus monkeys. Behaviour 85: 224–241.

Desmond (1994) Huxley: from devil's disciple to evolution's high priest. New York: Perseus.

Dewey J (1993 [1898]) Evolution and ethics. Reprinted in: Nitecki MH, Nitecki DV (eds) Evolutionary ethics. Albany, NY: State University of New York Press, pp. 95–110.

di Pellegrino G, Fadiga L, Fogassi L, Gallese V, Rizzolatti G (1992) Understanding motor events: A neurophysiological study. Exp Brain Res 91: 176–180.

Dimberg U (1982) Facial reactions to facial expressions. Psychophysiology 19: 643–647.

Dimberg U (1990) Facial electromyographic reactions and autonomic activity to auditory stimuli. Biol Psychol 31: 137–147.

Dimberg U, Thunberg M Elmehed K (2000) Unconscious facial reactions to emotional facial expressions. Psychol Scie 11: 86–89.

Eisenberg N (2000) Empathy and sympathy. In: Lewis M, Haviland-Jones JM (eds) Handbook of emotion. (2nd ed). New York: Guilford Press, pp 677–691

Fehr E, Schmidt KM (1999) A theory of fairness, competition, and cooperation. Quart J Econ 114: 817–868.

Flack JC, de Waal FBM (2000) 'Any animal whatever:' Darwinian building blocks of morality in monkeys and apes. J Consciousness Stud 7 (1–2): 1–29.

Frank R.H (1988) Passions within reason: the strategic role of the emotions. New York: Norton.

Freud S (1930) Civilization and its discontents. New York: Norton.

Gallese V (2001) The 'shared manifold' hypothesis: From mirror neurons to empathy. In: Thompson E (ed) Between ourselves: Second-person issues in the study of consciousness. Thorverton, UK: Imprint Academic, pp 33–50

Gallup GG (1982) Self-awareness and the emergence of mind in primates. Am J Primatol 2: 237–248.

Ghiselin M (1974) The economy of nature and the evolution of sex. Berkeley, CA: University of California Press.

Gould SJ (1980) So cleverly kind an animal. In: Ever since DarwinHarmondsworth, UK: Penguin, pp 260–267

Greene J, Haidt J (2002) How (and where) does moral judgement work? Trends Cogn Sci 16: 517–523.

Greene J (2005) Emotion and Cognition in moral judgement: evidence from neuroimaging. In: Changeux JP, Damasio A, Singer W, Christen Y (eds) Neurobiology of human values. Heidelberg. Springer Verlag.

Haidt J (2001) The emotional dog and its rational tail: A social intuitionist approach to moral judgment. Psychol Rev 108: 814–834.

Hatfield E, Cacioppo JT, Rapson RL (1993) Emotional contagion. Curr Directions Psychol Sci 2: 96–99.

Hauser MD (2000) Wild minds: what animals really think. New York: Holt.

Hirschleifer J (1987) On the emotions as guarantors of threats and promises. In: Dupre J (ed) The latest on the best: essays in evolution and optimality. Cambridge, MA: MIT Press, pp 307–326

Hobbes T (1991 [1651]) Leviathan. Cambridge: Cambridge University Press.

Hoffman ML (1981) Perspectives on the difference between understanding People and understanding Ttings: the Role of affect. In: Flavell JH, Ross L (eds) Social cognitive development. Cambridge: Cambridge University Press, pp 67–81

Hume D (1978 [1739]) A Ttreatise of human nature. Oxford : Oxford University Press.

Huxley TH. (1989 [1894]) Evolution and Ethics. Princeton, NJ: Princeton University Press.

Ladygina-Kohts NN (1935 [2001]) Infant chimpanzee and human child: a classic 1935 comparative study of ape emotions and intelligence. de Waal FMB (ed) New York: Oxford University Press.

Levenson RW, Reuf AM (1992) Empathy: a physiological substrate. J Personal Soc Psychol 63: 234–246.

Lipps T (1903) Einfühlung, innere Nachahmung und Organempfindung. Arch gesamte Psychol 1: 465–519.

Lorenz KZ (1963 [1966]) On aggression. London: Methuen.

Masserman J, Wechkin MS, Terris W (1964) Altruistic behavior in rhesus monkeys. Am J Psychiat 121: 584–585.

Miller RE, Murphy JV, Mirsky IA (1959) Relevance of facial expression and posture as cues in communication of affect between monkeys. AMA Arch Gen Psychiat 1: 480–488.

O'Connell SM (1995) Empathy in chimpanzees: Evidence for theory of mind? Primates 36: 397–410.

Povinelli DJ (1998) Can animals empathize? Maybe not. Sci Am http://geowords.com/lostlinks/b36/7.htm

Preston SD, de Waal FBM (2002a) Empathy: its ultimate and proximate bases. BehavBrain Sci 25: 1–72.

Preston SD, de Waal FBM (2002b) The communication of emotions and the possibility of empathy in animals. In: Post SG, Underwood LG, Schloss JP, Hurlbut WB (eds) Altruistic love: science, philosophy, and religion in dialogue. Oxford: Oxford University Press, pp 284–308

Preston SD, Bechara A, Grabowski TJ, Damasio H, Damasio AR (2002) Functional anatomy of emotional imagery: Positron Emission Tomography of personal and hypothetical experiences. J Cogn Neurosci Suppl 14, 126.

Prinz W, Hommel B (2002) Common mechanisms in perception and action. Oxford: Oxford University Press.

Reiss D, Marino L (2001) Mirror self-recognition in the bottlenose dolphin: A case of cognitive convergence. Proc Natl Acad Sci USA 98: 5937–5942.

Roes F (1997) An interview of Richard Dawkins. Human Ethol Bull 12: 1–3.

Sanfey AG, Rilling JK. Aronson JA, Nystrom LE, Cohen JD (2003) The neural basis of economic decision-making in the Ultimatum game. Science 300: 1755–1758.

Schino G (2000) Beyond the primates: expanding the reconciliation horizon. In: Aureli F, de Waal FMB (eds) Natural conflict resolution. Berkeley, CA: University of California Press, pp 225–242

Singer T, Seymour B, O'Doherty J. Kaube H, Dolan RJ, Frith CD (2004) Empathy for pain involves the affective but not sensory components of pain. Science 303: 1157–1162.

Smith A (1937 [1759]) A theory of moral sentiments. New York: Modern Library.

Trivers RL (1971) The evolution of reciprocal altruism. Quart Rev Biol 46: 35–57.

Verbeek P, de Waal FBM (2001) Peacemaking among preschool children. Peace and conflict: J Peace Psychol 7: 5–28.

Verbeek P, Hartup WW, Collins WC (2000) Conflict management in children and adolescents. In: Aureli F, de Waal FBM (eds) Natural conflict resolution. Berkeley, CA: University of California Press, pp 34–53

Watts D.P, Colmenares F, Arnold K (2000) Redirection, consolation, and male policing: How targets of aggression interact with bystanders. In: Aureli F, de Waal FMB (eds) Natural conflict resolution. Berkeley, CA: University of California Press, pp 281–301

Westermarck E (1912 [1906]) The origin and development of the moral ideas. Volume 1, 2nd Edition. London: Macmillan.

Westermarck E (1917 [1908]) The origin and development of the moral ideas. Volume 2, 2nd Edition. London: Macmillan.

Wechkin S, Masserman JH, Terris Jr W (1964) Shock to a conspecific as an aversive stimulus. Psychonomic Sci 1:47–48

Wicker B, Keysers C, Plailly J, Royet JP, Gallese V, Rizzolatti G (2003) Both of us disgusted in My insula: the common neural basis of seeing and feeling disgust. Neuron 40:655–64

Williams GC (1988) Reply to comments on "Huxley's evolution and ethics in sociobiological perspective." Zygon 23: 437–438.

Williams JHG, Whiten A, Suddendorf T, Perrett DI (2001) Imitation, mirror neurons and autism. Neurosci Biobehav Rev 25: 287–295.

Wispé L (1991) The psychology of Sympathy. New York : Plenum.

Wolpert DM, Ghahramani Z, Flanagan JR (2001) Perspectives and problems in motor learning. Trends Cogn Sci 5: 487–494.

Wright R (1994) The moral animal: the new science of evolutionary psychology. New York: Pantheon.

Yerkes RM (1925) Almost human. New York: Century.
Zahn-Waxler C, Hollenbeck B, Radke-Yarrow M (1984) The origins of empathy and altruism. In: Fox MW, Mickley LD (eds) Advances in animal welfare science. Washington, DC: Humane Society of the United States, pp 21–39
Zahn-Waxler C, Radke-Yarrow M, Wagner E, Chapman M (1992) Development of concern for others. Dev Psychol 28: 126-136.

Disorders of Social Conduct Following Damage to Prefrontal Cortices

Hanna Damasio[1]

Some of the most important human values pertain directly to social conduct, and it is apparent that the study of disorders of social conduct from a neurological perspective can yield important information regarding the neurobiology of human values. My purpose in this chapter is to review the evidence I regard as most relevant in this area.

Although it is manifestly difficult to define "normal" social conduct, most readers would agree on what generally constitutes good or bad social conduct. I will assume that agreement and simply say that normal social conduct is that which respects 1) the social conventions of a given culture, and 2) the ethical rules and laws of that culture.

I also assume that most readers expect normal social conduct to result not only from the acquisition of socio-cultural knowledge and skills, but also from biological (and especially neurobiological) factors. In fact, I suspect most believe that an interaction of socio-cultural and biological factors is necessary for normal social conduct to emerge, and would agree that in order for normal social conduct to occur, certain conditions have to be met: 1) the biological devices necessary for the behavior must have been put in place by the genome; 2) the current environment must be compatible with the behavior; 3) the current biological state of the individual's organism must be such that the biological devices are operational; and 4) the past environment must have allowed the individual to develop biologically and psychologically in such a way that the relation between the behavior and the social situation could be properly adjusted.

Once these assumptions are agreed upon, the question I would like to pose is as follows: what do we know about the biological devices necessary for normal social conduct to occur?

A first line of evidence that helps answer this question comes from lesion studies in neurological patients. Here I will present results from the investigation of a large group of patients who have had a neurological event that damaged their frontal lobes. In some, the damage occurs in adulthood; more rarely, it can also occur in childhood. The critical event is often the resection of a brain tumor (e.g., a meningeoma) but can be a cerebrovascular accident (e.g., a ruptured anterior communicating aneurism; a stroke in the territory of the anterior cerebral artery) or even head trauma. The principal damage centers on the ventral and mesial sectors of the prefrontal cortices (Fig. 1).

[1] University of Iowa College of Medicine, Department Neurology, Iowa City IA 52242, USA; e-mail: hanna-damasio@uiowa.edu

Changeux et al.
Neurobiology of Human Values
© Springer-Verlag Berlin Heidelberg 2005

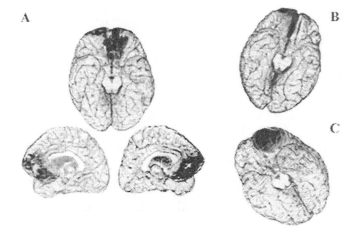

Fig. 1. Three-dimensional reconstruction, from high resolution MR scans, of 3 subjects, fully right-handed, with adult-onset ventro-medial prefrontal damage. In **A** the damage is bilateral, the result of resection of a meningeoma, in a 54 year old woman; in **B** the damage only involves the right ventro-medial prefrontal region and results from an infarct in the territory of the anterior cerebral artery, in a 61 year old man; and in **C** the damage is limited to the left anterior and medial ventral prefrontal region, the result of resection of a cystic lesion in a 32 year old woman.

Adult-onset frontal damage

In the first set of cases, I shall discuss damage acquired in adulthood. The typical patient is someone who underwent normal development and had normal social conduct prior to the onset of neurological disease. Because the patient's behavior was normal prior to the onset of brain dysfunction, and because the environment did not change before or after the onset of abnormal behaviors, this type of patient allows us to investigate abnormal behaviors that cannot be explained by genomic, biodevelopmental, sociodevelopmental, or cultural factors.

These patients have been described in several previous publications (Damasio et al. 1991; Damasio 1994) and I will merely review the highlights of their presentation. They have normal general intelligence; their sensory and motor skills are unchanged, and so are conventional memory, speech and language. However, their daily conduct has changed radically after the neurological event. The patients no longer keep their commitments; they do not show up on time for their jobs and do not observe the steps necessary to complete a task; they are derailed by irrelevant side issues instead; they cannot make plans for their immediate or distant future. They show a flattening of their primary emotions, and their social emotions – such as embarrassment, shame, compassion – are severely impaired.

When they are studied in the laboratory, these patients present with normal profiles in the basic neuropsychological tasks (verbal and performance IQ, learning and memory. language, reasoning skills). Even in special tasks measuring the ability to make cognitive estimates and to judge recency and frequency, or in

The Wisconsin Card Sorting Task, all of which are usually associated with frontal lobe dysfunction, these patients tend to be normal. This is also the case for tasks measuring more specifically the capacity for social problem-solving, such as the Optional Thinking Test (Platt and Spivack 1977; Spivack et al. 1976), designed to measure the ability to generate alternative solutions to a social dilemma, the Awareness of Consequences Test (Platt and Spivack 1977; Spivack et al. 1976), designed to measure the ability to generate a list of consequences of a particular action, and the Means-Ends Problem-Solving Procedure (Platt and Spivack 1974, 1977), which measures the ability to conceptualize step-by-step means to achieve a certain goal. In brief, patients with adult-onset damage to ventromedial prefrontal cortices (VMPFC) behave in these tasks in the same way that normal adult subjects do. In a task designed to assess the ability to resolve social and moral situations, the Kohlberg paradigm of the Standard Moral Interview (Colby and Kohlberg 1987), five of six patients attained the second level, the level that is characteristic of most adults, and one patient reached the third level, a level only attained by a minority of adults (Saver and Damasio 1991). Once again, these behaviors are indistinguishable from those of a normal population. Without a doubt, adult-onset VMPFC patients exhibit abnormal social problem solving in real life and real time but show normal social problem solving when they are tested in a laboratory setting.

The situation changes dramatically, however, when these patients are exposed to emotionally competent stimuli, such as familiar faces or scenes depicting suffering: they show abnormally low skin conductance responses (SCRs), even while they recognize the unique faces normally and correctly describe the situations of suffering.

And these patients' behavior is equally abnormal in another laboratory task designed to test decision-making under ambiguity, in a complex setting that resembles real life in that it involves rewards and punishments encountered under uncertainty: The Iowa Gambling Task (Bechara et al. 1994). The Iowa Gambling Task (IGT) was developed to investigate and measure the disability in decision-making that VMPFC patients show in real life. Here I will briefly call attention to its most important features.

The task involves selections of cards from four decks (A, B, C, and D), one card at a time. The subject is told that, after each card selection, there will be a monetary gain and that, from time to time, there will also be a monetary penalty. The subject is given an initial loan of $2,000.00 and told that the aim of the task is to make as much money as possible. What the subject is not told is that
1) the rewards in A and B are always $100, whereas in C and D they are only $50;
2) the unexpected penalties in A and B are high, up to $1,200, whereas in C and D they are small (the maximum being $300);
3) the game lasts for 100 card turns; and
4) turning cards from A and B leads to final losses in spite of immediate high rewards, whereas turning cards from C and D provides a slow but persistent accrual of gains.

Initially, both normal subjects and VMPFC patients tend to sample from all decks. When either normals or patients encounter high penalties, they tend to switch away from the high reward, high penalty decks. While both normals and patients do revisit those "bad" decks, VMPFC patients tend to revisit them more often than normal subjects do. In the end, normal subjects turn more cards from decks C and D (the winning, "good decks"), ending the task with a profit, whereas VMPFC patients turn more cards from decks A and B (the losing, "bad decks"), ending in bankruptcy (see Bechara et al. 1994). The IGT captures the abnormal decision-making these patients show in their daily lives. This abnormality is probably due to the complexity and uncertainty of the task. There are clear conflicts between immediate and long-term gains. Only continued performance reveals the best strategy, not unlike life itself.

The IGT can be played while skin conductance responses (SCRs) are continually monitored so that different stages of the performance can be analyzed separately from the standpoint of psychophysiology: the SCRs that occur immediately after a punishment or reward (we call this the "consequent" SCR), and the SCRs that occur in the 5 sec interval preceding the actual choice of a card (the "anticipatory" SCR). Both normals and patients show consequent SCRs, which tend to be of higher amplitude after a punishment than after a reward. On average, the amplitude of the responses in VMPFC patients is slightly lower than that of normal subjects, but the ranges overlap. With respect to the anticipatory SCRs, the two groups diverge remarkably. Normal subjects show anticipatory SCRs that discriminate between the decks: the responses are of high amplitude for decks A and B (the bad decks), and low amplitude for decks C and D (the good decks). Furthermore, the amplitude of the anticipatory SCRs to the bad decks starts to increase early on, when the subjects are still sampling from all decks. VMPFC patients are completely different: their anticipatory SCRs are of uniformly very low amplitude for *all decks*, thus failing to show any discrimination between good and bad decks. SCRs are an index, albeit incomplete, measure of the emotional responses of the subject, thus capturing the emotional defect of these patients.

Our interpretation of the IGT results, which draws on the Somatic Marker Hypothesis (Damasio et al. 1991; Damasio 1994, 1996), has been that the absence of a "somatic marker," which is evidenced by the lack of anticipatory SCRs, precludes an advantageous performance, the preference for decks C and D seen in normal subjects, even in those VMPFC subjects who eventually realize, as many but not all normal subjects do, the "badness" and "goodness" of the different decks. In brief, given the complex conflicts between reward and punishment over time that are present in the task, we believe that normal subjects must rely not only on their cold analysis of the situation but also on an emotionally related signal that steers their reasoning and choice to the good decks. This signal, the somatic marker, which can be conscious as in a gut feeling or non-conscious, is related to their prior experience of comparable situations of conflict.

Another explanation that has been adduced for the patients' behavior in the IGT is that they fail to learn that the contingencies are being reversed, that is, that a high-reward deck suddenly becomes a high-penalty deck. On that explanation patients would simply stick, perseveratively, to the bad deck. This explanation is not satisfactory. Many patients with prefrontal damage in the ventromedial sector (as opposed to the dorsolateral sector or lateral orbital sectors) who show the neuropsychological profile described earlier exhibit normal performance in tasks measuring the ability to learn contingency reversals. This is also evident in their ability to perform well in the Wisconsin Card Sorting Task (see also Busemeyer and Stout 2002; Stout et al. 2002).

Patients with dorsolateral prefrontal damage may also have abnormal social conduct and also fail the IGT. However, they also show overt cognitive impairments. Their social conduct and decision-making defect are part of a larger picture of cognitive defects. The dissociation between general preserved intellect and impaired decision-making that characterized the VMPFC patients does not obtain for the DLPFC patients. The emotional impairments are not as distinct (see Fig. 2).

A subgroup of patients with VMPFC damage, those in whom the lesion involves the posterior sector of the mesial orbital region and damages the basal forebrain area, may show defects in tasks of contingency reversal learning (Bechara et al. 2003; Bechara and Damasio 2004). Such defects are seen, for instance, in patients with damage due to rupture of anterior communicating aneurysms (Fig. 3; see also images in Fellows and Farah 2003).

Early-onset frontal damage

We have also studied patients with lesions similar to those described in the adult-onset group but acquired early in life, in fact as early as the first day of life and as late as age seven. In the 13 cases we have studied so far, we have investigated

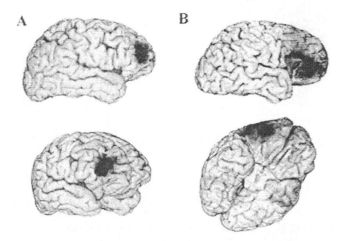

Fig. 2.   Three-dimensional reconstruction, from high resolution MR scans, of two right-handed adults with damage to the dorso-lateral sector of the right prefrontal cortex region. In A, a 54 year old man, the damage resulting from head trauma involves only dorso-lateral prefrontal regions; in B, a 22 year old woman, the damage involves the lower aspect of the right dorsolateral region and extends into the lateral ventral section, the result of a surgical ablation for the treatment of seizures.

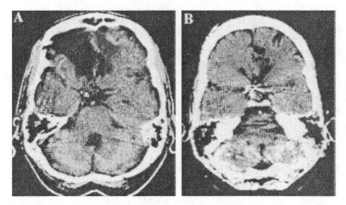

Fig. 3. Examples of two patients with VMPFC damage in **A** due to meningeoma resection, in **B** due to a ruptured anterior communicating aneurysm. Note that the lesion in **A** does not reach the posterior part of the ventro-medial sector whereas in **B** the lesion does involve the most posterior part of this territory.

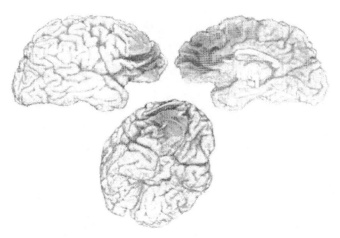

Fig. 4. Three-dimensional reconstruction, from a high resolution MR scan, of the brain of a 27 year old man, right-handed, showing extensive damage to the right prefrontal region, both ventral and medial, and dorsolateral as well. The damage was due to perinatal trauma.

most of these patients as young adults, long after the onset of their lesions. As in the adult-onset group, the cause is quite varied and the damage can be bilateral or unilateral (Fig. 4). And as with the adult-onset group, these patients are of normal intelligence and their sensory and motor skills, conventional memory, speech and language are normal too. As young children, they exhibit dysfunctional social interactions, both at school and at home. They show difficult behavior control and are insensitive to punishment. In spite of their normal intelligence, they usually need special schooling because of poor working habits. They do not make friends.

After high school graduation, once they lose a relatively structured environment, their social and behavior problems worsen remarkably. They never make plans for the future, do not seek employment, and cannot maintain the jobs that may be procured for them. Stealing is a common problem from early on and they often get into legal problems (Anderson et al. 1999).

The neuropsychological profile of these patients is normal, similar to what happens in the adult-onset group, but they too show hypo-emotionality and a remarkable absence of social emotions, and their IGT is abnormal. The remarkable difference, relative to the adult-onset group, appears in the results of tasks measuring social interactions. Here the early-onset VMPFC patients perform *abnormally* on all the tasks in which the adult-onset patients do so well (the Optional Thinking test, the Awareness of Consequences test, the Means-Ends Problem-Solving procedure and the Standard Moral Interview). Of note, in the Kohlberg paradigm, they do not advance beyond the first level, which is the level most normal children reach by age nine (see Anderson et al. 1999).

In brief, adult-onset VMPFC patients have abnormal social conduct but appear to know the rules they violate, whereas early-onset VMPFC patients not only show abnormal social conduct but also seem not to have learned the social rules that govern that social conduct.

Neuroanatomical and functional conclusions

What are the salient anatomical findings related to abnormal social conduct? First, damage to VMPFC, bilaterally or unilaterally, acquired in adulthood or in childhood, causes a marked disturbance of social conduct in the absence of major cognitive deficits. The presence and strength of the defect probably depend on the side of the damage and on the gender of the subject. An ongoing study from our laboratory seems to suggest that damage in the right hemisphere is more likely to cause the defect in men than women, whereas left hemisphere damage has the reverse effect (for examples, see Fig. 1B and C).

Second, damage to the posterior sector of the VMPFC next to the basal forebrain region and damage to the dorsolateral prefrontal sector are also associated with problems of social conduct, but these are found in the setting of more widespread cognitive deficits. The same can be said about two other groups of patients, those with damage to non-dominant somatosensory cortices and those with bilateral amygdala damage. They show features of abnormal social conduct, but their social conduct deficits also appear in the setting of more widespread cognitive deficits.

Third, the most salient behavioral/cognitive finding in all of these patients seems to be that disorders of social conduct have an obligatory accompaniment: a disorder of emotion and feeling.

An interpretation

What is the role played by the VMPFC in this disorder? And why is an emotional impairment an obligatory accompaniment? The answers we propose are as follows.

The VMPFC is likely to play multiple roles in normals and its damage is thus likely to cause several defects. In all likelihood, the VMPFC region is necessary for triggering most of the salient social emotions, its functional role in this regard being akin to that of the amygdala regarding fear. In the absence of spontaneous social emotions, early-onset VMPFC patients would be deprived of normal interactions with others and would be the recipients of potentially abnormal reactions from others. It would be difficult for these young patients to acquire a normal view of social interactions given these dually perturbed relationships.

But the VMPFC is likely to play another role, that of a learned repository of the link between situations calling for a decision and the outcomes of decisions, on the one hand, and, on the other, the emotional state associated with the situations or with the result of the decisions. The VMPFC would house the record of the linkage between "situation-and-associated-emotion" or "outcome-and-resulting-emotion." When a situation of the same category as one previously experienced would present itself, the VMPFC would generate the emotion previously associated with that category of situation and with the outcome of the decision it prompted. In this role, the VMPFC is, once again, a trigger region for emotions, except that here the trigger is not an evolutionarily established, emotionally competent stimulus, such as the suffering of another, which triggers compassion. Rather, the trigger is a stimulus learned by the subject, in his or her past experience, and the triggered emotion may or may not be a social emotion.

Both in early-onset and adult-onset VMPFC lesions, this operation would be precluded. This would account for most of the defects seen in the adult-onset patients and for a good part of the defects seen in the early-onset cases.

It should be noted that the normal VMPFC would operate by acting on other brain structures, namely, the DLPFC where we suspect signaling from VMPFC plays a role in reasoning and decision-making, directly or via the intermediary action of subcortical regions, e.g., amygdala. In other words, the "somatic marker" signal evoked in VMPFC would depend on contributions from subcortical structures and exert its role in the guidance of behavior via the regions on which reasoning is likely to depend the most given their role in higher cognition: the DLPFC.

This interpretation also helps explain why emotion impairments are consistently present in these patients. First, the social emotions embody the sort of social knowledge without which normal social conduct cannot be shaped through gradual fine tuning in childhood and adolescence. Second, emotions, especially of the social variety, play a critical role in acquiring knowledge about which actions produce good and right outcomes or bad and wrong outcomes. This is because positive and negative emotions incorporate, respectively, elements of reward or punishment that qualify the actions in terms of the judgment of others and the individual's own experience. Moreover, in adult life, the acquired emotional knowledge regarding the link between actions and outcomes would play a role in the guidance of behavior. Emotions, in their many varieties, would be an indispensable tool in the construction and utilization of adequate social knowledge.

Patients with different kinds of pathology, namely, psychiatric conditions, offer evidence that converges with this interpretation. Both Cleckley (1955) and Hare (1993) have shown that psychopaths whose intelligence can be regarded as normal when evaluated with standard tests have a profound blunting of emotion, a profile somewhat reminiscent of that seen in patients with VMPFC damage. Furthermore, Hare and Quinn (1971) demonstrated that emotionally–competent stimuli fail to evoke SCRs in such individuals, supporting the notion that in these patients the threshold for the triggering of emotion is abnormally high. And more recently, Raine et al. (2000) have shown that criminal psychopaths, but not drug addicts and not non-psychopathic psychiatric patients, have a significant volumetric reduction of pre-frontal lobe structures.

I should note that there is no evidence that psychopathic criminals have structural lesions of the frontal lobe of the sort we have described in our studies. It is likely, however, that these prefrontal cortices malfunction for a variety of reasons. Those reasons include biodevelopmental defects associated with social culture, nutritional, genetic, and acquired diseases of the central nervous system. Conceivably, some of them might actually have sustained lesions similar to those in our cases. This is a matter for future investigations.

In conclusion, as we consider these findings together they suggest the following:

1) certain disorders of social conduct in which ethical rules are violated can be related to specific brain systems and accounted for by neural dysfunction in the absence of causative sociocultural factors;
2) abnormal emotional function seems to play a critical role in disorders of social conduct.

References

Anderson WW, Bechara A, Damasio H, Tranel D, Damasio AR (1999) Impairment of social and moral behavior related to early damage in the human prefrontal cortex. Nature Neurosci 2: 1032–1037

Bechara A (2003) Risky business: Emotion, decision-making and addiction. J Gambling Stud 19: 23–51

Bechara A, Damasio A (2004) The somatic marker hypothesis: a neural theory of economic decision. Games Econ Behav 1 (special issue on neuroscience economics): 1–37

Bechara A, Damasio AR, Damasio H, Anderson SW (1994) Insensitivity to future consequences following damage to human prefrontal cortex. Cognition 50: 7–15

Bechara A, Tranel D, Damasio H, Damasio AR (1996) Failure to respond autonomically to anticipated future outcomes following damage to prefrontal cortex. Cereb Cortex 6: 215–225

Bechara A, Damasio H, Tranel D, Anderson SW (1998) Dissociation of working memory from decision making within the human prefrontal cortex. J Neurosci 18: 428–437

Bechara A, Damasio H, Damasio AR (2003) The role of the amygdala in decision-making. Ann NY Acad Sci 985: 356–369

Busemeyer JR, Stout JC (2002) A contribution of cognitive decision models to clinical assessment: decomposing performance on the Bechara gambling task. Psychol Assessment 14: 253–262

Cleckley H (1955) The mask of sanity. St. Louis, Mo: CV Mosby

Colby A, Kohlberg L (1987) The measurement of moral judgment. New York: Cambridge University Press

Damasio AR (1994) Descartes' error: emotion, reason, and the human brain. Grosset/Putnam

Damasio AR (1996) The somatic marker hypothesis and the possible functions of the prefrontal cortex. Phil Trans Royal Soc London (Biol) 351: 1413–1420

Damasio AR, Tranel D, Damasio H (1991) Somatic markers and the guidance of behavior: theory and preliminary testing. In: Levin HS, Eisenberg HM, Benton AL (eds) Frontal lobe function and dysfunction. Oxford University Press, pp 217–229

Fellows LK, Farah MJ (2003) Ventromedial frontal cortex mediates affective shifting in humans: evidence from a reversal learning paradigm. Brain 126: 1830–1837

Hare RD (1993) Without conscience. New York: Pocket Books

Hare RD, Quinn MJ (1971) Psychopathy and autonomic conditioning. J Abnormal Psychol77: 223–235

Platt J, Spivack G (1974) Means of solving real-life problems: I. Psychiatric patients versus controls and cross-cultural comparisons of normal females. J Commun Psychol 2:45–48

Platt J, Spivack G (1977) Measures of interpersonal problem-solving for adults and adolescents. Philadelphia: Department of Mental Health Sciences, Hahnemann Medical College

Platt G, Platt JJ, Shure MB (1976) The problem-solving approach to adjustment. San Francisco: Jossey-Bass Publishers

Raine A, Lencz T, Bihrle S, LaCasse L, Colletti P (2000) Reduced prefrontal gray matter volume and reduced autonomic activity in antisocial personality disorder. Arch Gen Psychiat 57: 119–127

Saver JL, Damasio AR (1991) Preserved access and processing of social knowledge in a patient with acquired sociopathy due to ventromedial frontal damage. Neuropsychologia 29: 1241–1249

Spivack G, Platt JJ, Shure MB (1976) The problem-solvin approach to adjustment. San Francisco: Jossey-Bass.

Stout J, Busemeyer J, Bechara A, Lin A (2002) Cognitive modeling of decision making in a simulated gambling task in frontal or somatosensory cortex damage. J Cogn Neuroscie C27 Suppl. 75

The Neurobiological Grounding of Human Values

Antonio Damasio[1]

Human Values: The Issue of Origins

How do we humans develop the values that permit us to classify objects as beautiful or ugly and to judge actions as good or evil? What is the basis for the moral
judgments we pronounce? Where are good social conduct and ethical principles
grounded? Humans have long been preoccupied with such questions informally,
in their day-to-day lives, and quite formally within philosophy and anthropology.
Recently, however, as cognitive science and neurobiology have succeeded in ap-
proaching a number of mind and behavior problems, these questions have begun
to be asked from the cognitive and neurobiological perspectives as well.

A traditional answer to these questions assumes that there has been a historical process of values construction permitted by the extraordinary development of
human intelligence. The intelligent constructions would have been perfected and
transmitted through generations all the way to ourselves. Our aim here is to consider the degree of invention involved in that intelligent construction. Does the
construction really spring forth from the perception of human interactions and
the creative reasoning over such perceptions, with little or no antecedent in humans who would not have had human values and in nonhuman species? Or might
it be that the intelligent construction resembles somewhat more a discovery of
antecedents? The antecedents would have been present in biological structures
long before humans acknowledged their presence and began manipulating them
intelligently. The discovery would have been followed by abundant elaboration in
a social space.

We favor the latter possibility. We believe that there was a biological blueprint
for the intelligent construction of human values, and that the biological blueprint
was present in nonhuman species and early humans. We also believe that a variety of natural modes of biological response, which include those known as emotions, already embody such values. They too were present in nonhuman species
and early humans.

As will be made clear in the text ahead, we do not wish to minimize the role
of social interactions and cultural history in the construction, refinement, codi-
fication and transmission of those values. We are not reducing human values to
biological inherited instincts. We simply wish to suggest that the construction

[1] University of Iowa College of Medicine, Department Neurology, Iowa City IA 52242, USA;
e-mail: antonio-damasio@uiowa.edu

Changeux et al.
Neurobiology of Human Values
© Springer-Verlag Berlin Heidelberg 2005

was constrained and oriented in certain directions by preexisting biological conditions. It did not enjoy infinite freedom. In no way does this view reduce the merit of the intelligent construction; neither does it oblige culture to follow biology blindly.

The biological blueprint for human values can be found in the machinery of homeostasis

In all life forms, there is a collection of systems that permits organisms to maintain biological processes within the range compatible with life. In complex species, the regulation of life depends on a close interaction between brain systems and body-proper systems, and it is, in effect, controlled by a specific collection of well-coordinated brain regions (see Damasio 1999 and 2003 for a review).

Life regulation is anything but a neutral process. It is through-and-through an active, committed, struggling process that seeks optimal parameters for the continuation of life. It involves choices and preferences, although at the basic levels those choices and preferences are automatic. The life regulation system is built to achieve certain goals, among them the maintenance of health, the prevention of circumstances leading to death, and the procurement of states of life tending toward optimal function rather than merely neutral or defective function. In other words, the life regulation system – homeostasis, for short – inherently embodies values in the sense that it rejects certain conditions of operation, those that would lead to disease and death, and seeks conditions that lead to survival in optimal conditions. Homeostasis has clear preferences, likes and dislikes, if you will.

The levers of homeostasis are defined by conditions that conscious and reflective humans can easily describe as states of pain and punishment, at one end of the spectrum, or pleasure and reward at the other. What we label as pain and pleasure is, in effect, the experience of particular configurations of the physiological state characterized by certain chemical parameters of the internal milieu, by the smooth muscle tone of viscera, by behaviors enacted in the musculoskeletal system, and by the distribution of neuromodulators in neural tissue. States of pain and punishment, if maintained over long periods of time without counteraction, lead to disease and death; states of pleasure and reward lead to health and well-being.

It is a demonstrable fact that what we usually call *good* and *evil* is aligned with categories of actions related to particular ranges of homeostatic regulation. What we call good actions are, in general, those actions that lead to health and well-being states in an individual, a group or even a species. What we call evil, on the other hand, pertains to malaise, disease or death in the individual, the group, or the species.

The same lines of thinking can be applied to the origins of our ability to classify objects or situations as beautiful or ugly. When we consider the range of operation of homeostatic processes, we can objectively describe states of efficiency, states of inefficiency and states in between. Efficient regulatory states are those, for example, in which the performance of regulation is not only adequate but timely, with minimal consumption of energy, minimal impediment, ease, and smoothness. Given the multi-tiered structure of the homeostatic process, the notion of harmony is perfectly apt to describe such states. The inefficient part of the regulatory spectrum is characterized by higher energy consumption, inadequate and untimely performance, impediment, difficulty, raggedness, and discoordination. The notions of disharmony and discord are not far behind either.

We suggest that, at the dawn of human values, the objects we classified as beautiful were associated with efficient states, either because they occurred in life circumstances in which the homeostatic range was efficient or because, given the perceptual and motor design of an organism, the objects themselves were capable of causing efficient homeostatic states. By contrast, ugly objects were those associated with inefficient ranges of homeostasis, either by co-occurrence or by actual causation.

A cautionary note, comparable to the one we made earlier regarding the origins of the notions of good and evil, applies here as well. We are not trying to reduce esthetic perception to efficient or inefficient physiological states. There is more to the "sublime percept" we experience when we hear Bach or contemplate a canvas by Rembrandt. Without a doubt, there is an emotive and feeling state but there is also a particular kind of cognition associated with it, one that allows a highly coordinated, coherent, and rapid evocation of pertinent knowledge whose access is, in and of itself, the prelude to yet another wave of emotion and feeling. Again our intent here is not reductive. We are simply pointing to origins, to the moorings of the construction. We are not saying that the construction is the same as the moorings (see Changeux, 2005; Singer 2005 for related discussions).

Some natural modes of biological response embody human values

It is apparent that emotions – from the simple background emotions to the primary emotions such as happiness and sadness, fear and anger, surprise and disgust, as well as the more elaborate social emotions – are related to states of homeostatic regulation and bring together action programs that assist homeostatic regulation directly or indirectly. All emotions can play an important role in social processes, intervening to alert individuals to impending problems, to correct possible problems, or to reward effective solutions. These roles are especially apparent for the social emotions and their ensuing feelings. We are referring here to a large group of emotions, among which the prominent exemplars are compassion or sympathy, an emotion related to a concern for others that commonly results in feelings of empathy; the emotions of embarrassment, shame, and guilt, all concerned in one way or another with the blaming of the self for some action that violated social norms; the emotions of disgust, indignation, and contempt, all of which concern the blaming of others for a norm violation; and a remarkable and often forgotten group of emotions that include gratitude, awe and admiration, pride, and elevation, all of which relate to praise for others or for the self as a result of perceiving some highly efficient solution to a particular problem, one that tends to coincide with respect for a highly-prized set of social norms (see Haidt 2001, 2002 for reviews on the scope of social emotions).

Like all other emotions, social emotions are action programs, that is, packages of actions with a given pattern that succeed in modifying the state of the emoting organism and change the environment as well. Experiencing the social emotions results in social feelings and, as is the case with other emotions, leads to the acquisition of a baggage of related ideas, often referred to as "scripts," ideas that co-occur with the deployment of the social emotions. As is the case with other emotions, social emotions assist with learning, recall, and reasoning (Damasio 2003).

It is apparent from a consideration of the patterns of social emotions that they reflect profound human values that are commonly expressed in ethical and esthetic notions. For example, compassion is closely linked to kindness, forgiveness, generosity and the tendency to act in a comforting way; disgust, indignation and contempt are inseparable from the recognition of a violation of norms carried out by another, and the tendency to serve punishment on the violator; shame and guilt are inseparable from the recognition of self-committed violations and the tendency to punish the self preemptively; gratitude, awe, and elevation, all recognize, to different degrees, the noble (beautiful) action (or noble, beautiful object) along with the desire to reward the noble actor. Moral principles, including a budding system of justice, are patently embedded in these natural responses. So are some fundamental principles of esthetics.

It is also apparent that social emotions are present in nonhuman species. The great apes, some species of monkeys (such as the capuchins), wolves, bats, and birds demonstrate action programs that can be assimilated to those of human social emotions (de Waal 1996; Brosnan and de Waal 2003). This is especially clear with an emotion such as compassion, which appears quite fully fleshed in the great apes, but is also detectable in behaviors suggesting embarrassment, disgust and indignation, gratitude, and pride, found in numerous other species. We are not suggesting that these emotions are precisely the same in animals and humans, let alone the ensuing feelings. We are suggesting that a powerful precursor to those emotions is already present in nonhuman species.

How social emotions were biologically put together and became a fixture of so many nonhuman and human brains is not clear. Our suggestion is that they depended on the primitives provided by homeostatic regulation and by another set of related primitives concerned with both a drive for cooperation and a drive to serve punishments. The drive for cooperation is a means for delivering rewards to others and to the self, relative to actions conducive to the good and the beautiful, for individuals and for the community. The drive to serve punishments leads to pain, in the self and in others, as a consequence of actions that would have led to disruption of the homeostatic range in other individuals and, by extension, the group. Both the cooperative and the punisher drives would have been necessary to shape the sort of social emotions we developed (see Fehr and Gächter 2002; de Quervain et al, 2004). Other neural factors will have played a critical role as well. The expanded cognitive capacities permitted by the development of higher-order cortices in prefrontal, temporal and parietal regions are important examples (Fuster 1989; Dehaene 2005). And so is the arrival into the brain of the most complex species of a class of neurons known as "mirror neurons" (Rizzolatti and Craighero 2005).

From social emotions to human values

Elsewhere we have outlined the mechanisms required for an emotionally competent stimulus (ECS) to cause an emotional state and a feeling (Damasio 1994, 1999; Damasio et al. 2000). From the processing perspective, we have postulated four stages: 1) the appraisal of the ECS; 2) the triggering of the emotion; 3) the execution; and 4) the emotional state. Feeling of the emotional state follows. Although these stages occur largely in sequence, there is evidence that the process includes recursions and reiterations that add to its complexity (Adolphs et al. 2005; Rudrauf 2005).

The key neural structures involved in the process are:
1) the sensory association and higher-order cortices in which appraisal of the ECS occurs;
2) the amygdala, ventromedial prefrontal cortices and the anterior insula, in which the triggering takes place and whose activity also influences appraisal;
3) the basal forebrain, hypothalamus, brain stem nuclei, and anterior cingulate cortex, which are the direct executors of the emotional state; and
4) the varied compartments of the body proper (internal milieu, viscera, musculo-skeletal system) and central nervous system, in which the emotional state comes to be fully instantiated.

The feeling stage depends on a host of somatosensing regions in brain stem and somatosensory cortices and on higher-order cortices. The experience of body-proper aspects of feeling – whether actually implemented in the body or simulated in CNS – depends on the former brain regions. The aspects of feeling related to the evocation of ideas and scripts consonant with the emotion depends on the higher-order cortices. The diagrams in Figures 1–3 provide a quick summary of these mechanisms.

How could we have bridged the distance between social emotions and human values as we know them now? Here is one possibility. A suitable ECS – for example, the sight of an individual who is suffering or the witnessing of either violations or observances of established norms – would have been evaluated and triggered a social emotion and the corresponding feeling. A moral or esthetic intuition would have ensued. Over time, the intuition would have been culturally fine-tuned: debated by the collective, enhanced or diminished, or even suppressed. The products of this cultural fine-tuning would have been transmitted, first orally then by written records. Eventually, they would have been codified in the form of social conventions: ethical rules, laws and systems of justice, and esthetic canons.

Deploying Human Values and Practicing a Knowledge

How does the system we are outlining here work in practice? We will try to answer with an example drawn from the means and manner with which we are presumed to produce moral judgments. Two main traditions are usually identified on this issue. One claims that, when we are faced with a situation, we use reasoning to detect a violation or an observance of the established norm, we use further reason-

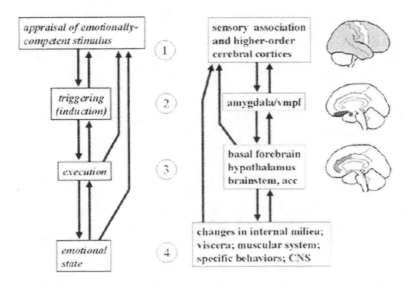

Fig. 1. Stages and structures in the emotional process

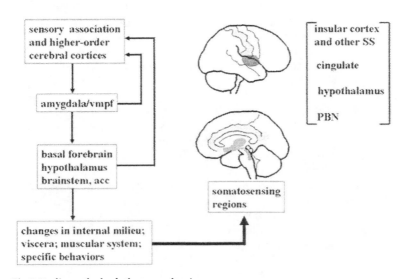

Fig 2. Feelings: the body-loop mechanism

ing to weigh and classify the violation or observance, and still more reasoning to pronounce a judgment and a sentence. This is the tradition identified with Kant, whose roots go back to Plato and whose modern exponent is the philosopher John Rawls. The psychologists Piaget and Kohlberg are also identified with this tradition.

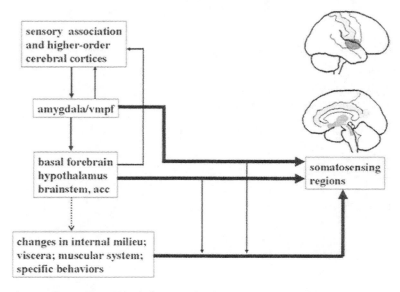

Fig 3. Feelings: the as-if-body-loop mechanism

The other tradition, which is identified with David Hume and Adam Smith, claims that we react to the social situation emotionally and automatically. We instantly produce moral sentiments and intuitions that guide us towards our response to the situation. In this tradition, much of the "moral reasoning" occurs *after* the moral intuitions have given us a first response. This late, after-the-intuition reasoning often takes the form of a rationalization, the post-hoc construction of a case for a certain intuition rather than the deduction that led into it. Using Haidt's account(2002), we can say that moral intuitions are the judges but those judges are not against having attorneys construct a justification for their pronouncements. Moral reasonings, on the other hand, handle the whole judicial process. The kinder and gentler tradition of moral practice provided by moral intuition has some roots in Aristotle and, in a roundabout way, in Spinoza. Darwin and Freud were early adopters and Johathan Haidt has argued this view persuasively.

At first glance, the findings from our studies on emotion recommend a preference for this latter tradition. However, we must note that the moral intuition view does not capture human behavior comprehensively either. In effect, depending on the situations and the nature of the norm, it is apparent that approaches compatible with either tradition may be adopted with advantage. Moreover, it is likely that in most circumstances mechanisms compatible with *both* traditions will be called into action and that those mechanisms are interactive. The fact that social feelings include the evocation of innate and learned scripts typical of the particular social emotion probably means that moral intuitions are accompanied by a deployment of knowledge that will facilitate reasonings occurring in parallel to intuitions. Note that we are not referring here to the post-hoc reasonings that are used to justify Kantian moral judgments. Figure 4 sketches out the idea.

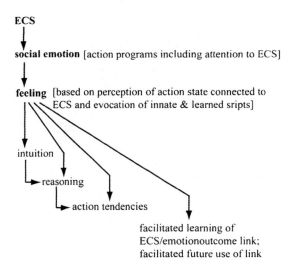

Fig 4. Social emotions and their consequences

In brief, we wish to make clear that, while we believe emotions, feelings and intuitions play a critical role in our use of human values, reasoning need not be excluded. It is perhaps most accurate to suggest that the balance between intuition and reasoning varies from case to case, and so does their interactivity. The role of emotions, feelings and intuitions is likely to be primary, with intuitions engaged first and reasoning following shortly thereafter. It is important, however, to avoid drawing an opposition between emotion and reasoning, and equally important not to oppose emotion to cognition. Emotions deliver ample cognitive information via feelings. We interpret the positions of Kahneman and Sunstein (2005) and Greene et al. (2001) on this issue to be similar to ours.

One final qualification pertains to the issue of the implicit knowledge covered under an intuitive process. Forceful probing for covert knowledge that does not become explicit during intuitions suggests that we may know far more than we believe we know as we reach an intuitive conclusion. The issue remains, however, that, if the knowledge is not available in consciousness during the actual unfolding of the intuitive process, it is of little or no practical use. It may also be the case that the careful and over-zealous probing for covert knowledge contributes to a partial fabrication of knowledge in a process not unlike that of post-hoc rationalization.

Neural Grounding

On the basis of recent neuroscience investigations, it appears that the neural systems necessary to support the knowledge of human values and to deploy such knowledge in the appropriate circumstances are as follows:
1)   brain systems necessary for social emotion;
2)   brain systems necessary for the experience of emotional feelings; and

3) brain systems that support the evocation of innate and acquired scripts related to the social emotions.

There is abundant evidence to indicate that the ventromedial sector of the prefrontal cortices (VMPFC) is a region necessary for the triggering of social emotions (Damasio 2003; Damasio 2005). This is a large territory, a portion of the even larger orbitofrontal cortices. The VMPFC sector is not only necessary for triggering several social emotions but is also related to the learning and recall of social knowledge. There is preliminary evidence (Bechara and Damasio 2004) that the anterior, middle, and posterior sectors of the VMPFC are functionally different, and the left and right halves of the system may be functionally distinct as well. The anterior insular cortices are likely to be yet another vital region for the triggering of social emotions, and it is possible that subcortical nuclei in the amygdala and basal forebrain are as well. The execution of the social emotions is likely to recruit the subcortical machinery needed for the primary emotions and exert comparable actions in both the body proper and the CNS, albeit in substantially different patterns.

The brain systems necessary to experience the emotional feelings arising from social emotions constitute a critical part of the neural grounding of human values. As in the case of other emotional feelings, the key neural support for these states is the complex of somatosensing regions that includes subcortical nuclei such as the parabrachial nucleus and the somatosensory cortices, especially but not exclusively those of the insula (see Damasio et al, 2000; Damasio 2003). Recent evidence from an experimental study of compassion-empathy in humans supports this notion (Singer et al. 2004).

Finally, there are brain structures that direct the evocation of the innate and acquired scripts that accompany the social emotions and constitute an integral part of the experience of such emotions. Such structures are located in the higher-order prefrontal cortices, especially in the dorsolateral and polar sectors. We envision the actual evocation of the explicit facts and skills contained in the scripts to depend on many other sectors of the cerebral cortex, namely the sensory cortices of varied modalities and the motor-related cortices.

It should be apparent, then, that there is not one center or system dedicated to one particular human value, let alone to all values. The deployment of human values is a concerted effort that involves numerous brain structures, in several systems, and that proceeds in a highly parallel and recursive manner. As is the rule in the case of other complex kinds of knowledge and skill, the task is shared by many entities working together. And yet the relative specificity of the neural entities linked to human values, social cognition, and social behavior is quite apparent, given that there are many brain structures that have little or nothing to contribute to this endeavor.

References

Adolphs R, Gosselin F, Buchanan T W, Tranel D, Schyns P, Damasio AR (2005) A mechanism for impaired fear recognition after amygdale damage, Nature 433:68–72

Bechara A, Damasio A (2004) The somatic marker hypothesis: a neural theory of economic decision. Games and economic behavior. Vol. 1 (special issue on neuroscience economics) pp 1–37

Brosnan SF, de Waal FBM (2003) Monkeys reject unequal pay, Nature 425: 297–299.

Changeux JP (2005) Mental synthesis and aesthetic perception. In: Changeux JP, Damasio A, Singer W, Christen Y (eds) Neurobiology of human values. Heidelberg: Springer Verlag, pp. 1–10.

Damasio AR (1994) Descartes' error: emotion, reason, and the human brain. New York: Grosset/Putnam.

Damasio A (1999) The feeling of what happens, body and emotion in the making of consciousness. New York: Harcourt, Brace & Company

Damasio AR, Grabowski TJ, Bechara A, Damasio H, Ponto LLB, Parvizi J, Hichwa RD (2000) Subcortical and cortical brain activity during the feeling of self-generated emotions. Nature Neurosci 3:1049–1056

Damasio A (2003) Looking for Spinoza, Joy, sorrow, and the feeling brain. New York: Harcourt, Brace & Company

Damasio H (2005) Disorders of social conduct following damage to prefrontal cortices. In: Changeux JP, Damasio A, Damasio H, Singer W, Christen Y (eds) The neurobiology of human values. Heidelberg: Springer Verlag, pp. 37–45.

Dehaene S (2005). How a primate brain comes to know some mathematical truths. In: Changeux JP, Damasio A, Singer W, Christen Y (eds) Neurobiology of human values. Heidelberg: Springer Verlag, pp. 143–155.

de Quervain D, Fischbacher U, Treyer V, Schellhammer M, Schnyder U, Buck A, Fehr E (2004) The neural basis of altruistic punishment. Science 305: 1254–1258

de Waal FBM (1996) Good natured: the origins of right and wrong in humans and other animals. Cambridge, Massachusetts: Harvard University Press

Fehr E, Gächter S (2002) Altruistic punishment in humans, Nature 415: 137–140

Fuster J (1989) The prefrontal cortex: anatomy, physiology, and neuropsychology of the frontal lobe. 2nd Ed. New York: Raven Press

Greene JD, Sommerville RB, Nystrom LE, Darley JM, Cohen JD (2001) An fMRI investigation of emotional engagement in moral judgment, Science 293: 2105–2108

Haidt J (2001) The emotional dog and its rational tail: a social intuitionist approach to moral judgment. Psychol Rev 108: 814–834

Haidt J (2002) The moral emotions. In: Davidson RJ, Scherer K, Goldsmith HH (eds) Handbook of affective sciences. Oxford University Press, pp. 852–870

Kahneman D (2005) Cognitive psychology of moral intuitions. In: Changeux JP, Damasio A, Singer W, Christen Y (eds) Neurobiology of human values. Heidelberg: Springer Verlag, pp. 91–105.

Rizzolatti G (2005) Mirror neurons: a neural approach to empathy. In: Changeux JP, Damasio A, Singer W, Christen Y (eds) Neurobiology of human values. Heidelberg: Springer Verlag, pp. 107–123.

Rudrauf D (2005). Aspects of the dynamics of the human cerebral cortex during emotion and feeling. PhD dissertation, University of Iowa and Université Pierre et Marie Curie

Singer W (2005) How does the brain know when it is right ? In: Changeux JP, Damasio A, Singer W, Christen Y (eds) Neurobiology of human values. Heidelberg: Springer Verlag, pp. 125–135.

Singer T, Seymour B, O'Doherty J, Kaube H, Dolan RJ, Frith CD (2004) Empathy for pain involves the affective but not sensory components of pain. Science 303: 1157–1162

Emotion and Cognition in Moral Judgment: Evidence from Neuroimaging

Joshua Greene[1]

Summary

Traditional theories of moral psychology emphasize reasoning and "higher cognition," while more recent work emphasizes the role of emotion. In this article, I discuss neuroimaging data that support a theory of moral judgment according to which both "cognitive" and emotional processes play crucial and sometimes mutually competitive roles. These data suggest that brain regions associated with cognitive control (anterior cingulate cortex and dorsolateral prefrontal cortex) are recruited to resolve difficult moral dilemmas in which utilitarian values require "personal" moral violations, violations associated with increased activity in brain regions associated with emotion and social cognition (medial prefrontal cortex, superior temporal sulcus, posterior cingulate cortex, temporal poles, and the amygdala). We have also found that brain regions including the anterior dorsolateral prefrontal cortices predict inter-trial differences in moral judgment behavior, exhibiting greater activity for utilitarian judgments. I speculate that our emerging psychological and neuroscientific understanding of moral judgment may influence our responses to real-world moral dilemmas.

For many decades, the dominant models of moral judgment have been "cognitive," treating moral judgment primarily as a reasoning process, especially in mature adults (Kohlberg 1969). In recent years, however, much of social psychology has taken an "affective turn," focusing on unconscious processes and implicit attitudes (Bargh and Chartrand 1999). Moral psychology has moved in this direction as well, largely through the work of Paul Rozin (Rozin et al. 1999) and Jonathan Haidt (2001), who have argued forcefully for the importance of emotion in moral decision-making. Neuroscientist Antonio Damasio (1994, 2003) and colleagues (Adolphs 2003; Adolphs et al. 1998; Anderson et al, 1999; Bechara et al. 2000) have made similar points regarding the role of emotion in real-world human decision-making.

Within the last five to ten years, functional magnetic resonance imaging (fMRI) has emerged as a prominent tool for studying the neural bases of human behavior, including social behavior. In this article I will describe some of my own work (conducted in collaboration with Jonathan Cohen, John Darley, and Leigh Nystrom) using fMRI to study the neural bases of moral decision-making. This

[1] Department of Psychology, Princeton University, Green Hall, Washington Road, Princeton NJ 08544, USA; e-mail: jdgreene@princeton.edu

Changeux et al.
Neurobiology of Human Values
© Springer-Verlag Berlin Heidelberg 2005

research suggests a synthesis of the two research traditions mentioned above. More specifically, our work suggests that moral judgments are produced through a complex interplay between intuitive emotional processes and controlled "cognitive" processes.

Emotional Engagement in Moral Judgment

Consider the following moral dilemma, known as the *trolley* dilemma (Foot 1978; Thomson 1986):

A runaway trolley is headed for five people who will be killed if it proceeds on its present course. The only way to save these people is to hit a switch that will turn the trolley onto a side track where it will run over and kill one person instead of five. Is it okay to turn the trolley in order to save five people at the expense of one?

The consensus among philosophers (Fischer and Ravizza 1992), as well as people who have been tested experimentally (Greene et al. 2001, 2004; Petrinovich and O'Neill 1996; Petrinovich et al. 1993), is that it is morally acceptable to save five lives at the expense of one in this case. Next consider the *footbridge* dilemma (Thomson 1986):

As before, a runaway trolley threatens to kill five people, but this time you are standing next to a large stranger on a footbridge spanning the tracks, in between the oncoming trolley and the five people. The only way to save the five people is to push this stranger off the bridge and onto the tracks below. He will die as a result, but his body will stop the trolley from reaching the others. Is it okay to save the five people by pushing this stranger to his death?

Here the consensus is that it is not okay to save five lives at the expense of one (Fischer and Ravizza 1992; Greene et al. 2001, 2004; Petrinovich and O'Neill 1996; Petrinovich et al. 1993).

Why do people respond differently to these two cases? Philosophers have offered a variety of *normative* explanations for why these two cases are different. That is, they have assumed that our responses to these cases are correct, or at least reasonable, and have sought principles that *justify* treating these two cases differently (Fischer and Ravizza 1992). For example, one might suppose, following Kant (Kant 1959) and Aquinas (Aquinas 1988), that it is wrong to harm someone as a means to helping someone else. In the *footbridge* case, the proposed action involves literally using the person on the footbridge as a trolley-stopper, whereas in the *trolley* case the victim is to be harmed merely as a side effect. (Were the single person on the alternate track to magically disappear, we would be very pleased.) In response to this proposal, Thomson devised the *loop* case (1986). Here, the situation is similar to that of the *trolley* dilemma, but this time the single person is on a piece of track that branches off of the main track and then rejoins it at a point before the five people. In this case, if the person were not on the side track, the trolley would return to the main track and run over the five people. The consensus here is that it is morally acceptable to turn the trolley in this case, despite the fact that here, as in the *footbridge* case, a person will be used as a means.

There have been many such normative attempts to solve the trolley problem, but none of them has been terribly successful (Fischer and Ravizza 1992). My collaborators and I have proposed a partial and purely descriptive solution. We hypothesized that the thought of pushing someone to his death in an "up close and personal" manner (as in the *footbridge* dilemma) is more emotionally salient than the thought of bringing about similar consequences in a more impersonal way (e.g., by hitting a switch, as in the *trolley* dilemma). We proposed that this difference in emotional response explains why people respond so differently to these two cases.

The rationale for distinguishing between *personal* and *impersonal* forms of harm is largely evolutionary. "Up close and personal" violence has been around for a very long time, reaching back far into our primate lineage (Wrangham and Peterson 1996). Given that personal violence is evolutionarily ancient, predating our recently evolved human capacities for complex abstract reasoning, it should come as no surprise if we have innate responses to personal violence that are powerful, but rather primitive. That is, we might expect humans to have negative emotional responses to certain basic forms of interpersonal violence, where these responses evolved as a means of regulating the behavior of creatures who are capable of intentionally harming one another, but whose survival depends on cooperation and individual restraint (Sober and Wilson 1998; Trivers 1971). In contrast, when a harm is *impersonal*, it should fail to trigger this alarm-like emotional response, allowing people to respond in a more "cognitive" way, perhaps employing a cost-benefit analysis. As Josef Stalin once said, "A single death is a tragedy; a million deaths is a statistic." His remarks suggest that, when harmful actions are sufficiently impersonal, they fail to push our emotional buttons, despite their seriousness, and as a result we think about them in a more detached, actuarial fashion.

This hypothesis makes some strong predictions regarding what we should see going on in people's brains while they are responding to dilemmas involving personal vs. impersonal harm (henceforth "personal" and "impersonal" moral dilemmas). The contemplation of personal moral dilemmas like the *footbridge* case should produce increased neural activity in brain regions associated with emotional response and social cognition, whereas the contemplation of impersonal moral dilemmas like the *trolley* case should produce relatively greater activity in brain regions associated with "higher cognition."[1] This is what we found (Greene et al. 2001, 2004). Contemplation of personal moral dilemmas produced relatively greater activity in three emotion-related areas: the posterior cingulate

---

[1] It turns out that determining what makes a moral dilemma "personal" and "like the footbridge case" vs. "impersonal" and "like the trolley case" is no simple matter, and in many ways re-introduces the complexities associated with traditional attempts to solve the trolley problem. For the purposes of this essay, however, I'm happy to leave the personal/impersonal distinction as an intuitive one, in keeping with the evolutionary account given above. For the purposes of designing the brain-imaging experiment discussed below, however, my collaborators and I developed a more rigid set of criteria for distinguishing personal from impersonal moral violations (Greene et al. 2001). I no longer believe that these criteria are adequate. Improving these is a goal of ongoing research.

cortex (Maddock 1999), the medial prefrontal cortex (Phan et al. 2002), and the amygdala (Adolphs 2003; Phan et al. 2002). This effect was also observed in the superior temporal sulcus and the temporal pole, regions associated with various kinds of social cognition (Allison et al. 2000; Saxe et al. 2004). At the same time, contemplation of impersonal moral dilemmas produced relatively greater neural activity in two classically "cognitive" brain areas, the dorsolateral prefrontal cortex and inferior parietal lobe.

This hypothesis also makes a prediction regarding people's reaction times. According to the view I've sketched, people tend to have emotional responses to personal moral violations, responses that incline them to judge against performing those actions. That means that someone who judges a personal moral violation to be *appropriate* (e.g., someone who says it is okay to push the man off the bridge in the *footbridge* case) will most likely have to override an emotional response in order to do it. This overriding process will take time, and thus we would expect that "yes" answers would take longer than "no" answers in response to personal moral dilemmas like the *footbridge* case. At the same time, we have no reason to predict a difference in reaction time between "yes" and "no" answers in response to impersonal moral dilemmas like the *trolley* case, because there is, according to this model, no emotional response (or much less of one) to override in such cases. Here, too, the prediction has held. Trials in which the subject judged in favor of personal moral violations took significantly longer than trials in which the subject judged against them, but there was no comparable reaction time effect observed in response to impersonal moral violations (Greene et al. 2001, 2004).

Other neuroscientific research suggests that emotions play a prominent, if not dominant, role in moral judgment (Anderson et al. 1999; Blair 2001; Damasio 1994; Kiehl et al. 2001; Moll et al. 2001, 2002a,b).

Cognitive Conflict and Control in Moral Judgment

In a subsequent set of analyses, we subdivided the personal moral dilemmas into two categories on the basis of difficulty (i.e., based on reaction time). Consider the following moral dilemma (the *crying baby* dilemma):

It is wartime, and you and some of your fellow villagers are hiding from enemy soldiers in a basement. Your baby starts to cry, and you cover your baby's mouth to block the sound. If you remove your hand, your baby will cry loudly, the soldiers will hear, and they will find you and the others and kill everyone they find, including you and your baby. If you do not remove your hand, your baby will smother to death. Is it okay to smother your baby to death in order to save yourself and the other villagers?

This is a very difficult question. Different people give different answers, and nearly everyone takes a relatively long time. This response is in contrast to other personal moral dilemmas, such as the *infanticide* dilemma, in which a teenage girl must decide whether to kill her unwanted newborn. In response to this case, people (at least the ones we tested) quickly and unanimously say that this action is wrong.

What's going on in these two cases? My collaborators and I hypothesized as follows. In both cases, there is a prepotent, negative emotional response to the personal violation in question, killing one's own baby. In the *crying baby* case, however, a cost-benefit analysis strongly favors smothering the baby. After all, the baby is going to die no matter what, and so you have nothing to lose (in consequentialist terms) and much to gain by smothering it, awful as it is. In some people, the emotional response dominates, and these people say "no." In other people, this "cognitive," cost-benefit analysis wins out, and these people say "yes."

What does this model predict that we will see going on in people's brains when we compare cases like *crying baby* and *infanticide*? First, this model supposes that cases like *crying baby* involve an increased level of "response conflict," that is, conflict between competing representations for behavioral response. Thus, we should expect that difficult moral dilemmas like *crying baby* will produce increased activity in a brain region that is associated with response conflict, the anterior cingulate cortex (Botvinick et al. 2001). Second, according to our model, the crucial difference between cases like *crying baby* and cases like *infanticide* is that the former evoke strong "cognitive" responses that can effectively compete with a prepotent, emotional response. Thus, we should expect to see increased activity in classically "cognitive" brain areas when we compare cases like *crying baby* to cases like *infanticide*, despite the fact that difficult dilemmas like *crying baby* are personal moral dilemmas, which were previously associated with emotional response (Greene et al. 2001).

These two predictions have held (Greene et al. 2004). Comparing high-reaction-time personal moral dilemmas like *crying baby* to low-reaction-time personal moral dilemmas like *infanticide* revealed increased activity in the anterior cingulate cortex (conflict) as well as the anterior dorsolateral prefrontal cortex and the inferior parietal lobes, both classically "cognitive" brain regions.

Cases like *crying baby* are especially interesting because they allow us to directly compare the neural activity associated with competing moral philosophies. Consequentialists (or utilitarians9 such as John Stuart Mill (1998) believe that one should, in any given decision, do whatever will produce the best overall consequences. In contrast, deontologists such as Immanuel Kant (1959) believe that it is often wrong to do things that will produce the best possible consequences ("The ends don't justify the means."). According to our model, when people say "yes" to cases like the *crying baby* case (the utilitarian answer), it is because the "cognitive," cost-benefit analysis has successfully dominated the prepotent emotional response that drives people to say "no" (the non-utilitarian or deontological answer). If that model is right, then we should expect to see increased activity in the previously identified "cognitive" brain regions (dorsolateral prefrontal cortex and inferior parietal cortex) for the trials in which people say "yes" in response to cases like *crying baby*. This result is exactly what we found. In other words, people exhibit more "cognitive" activity when they give the utilitarian answer[2] (Fig. 1).

---

[2] It is worth nothing that no brain regions, including those implicated in emotion, exhibited the opposite effect. However, it is difficult to draw conclusions from negative neuroimaging results because current neuroimaging techniques, which track changes in blood flow, are relatively crude instruments for detecting patterns in neural function.

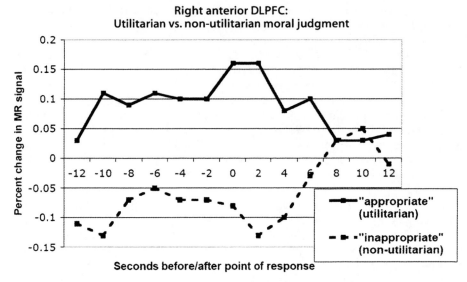

Fig. 1. Subjects responded to moral dilemmas in which someone must commit a "personal" moral violation if he is to bring about the best possible consequences, i.e., to act in accordance with utilitarian principles. Brain regions associated with cognitive control, the left and right anterior dorsolateral prefrontal cortices (DLPFC), exhibited increased activity during trials in which the subjects judged in accordance with utilitarianism. Above is a time course describing the average level of activity in a brain region within the right anterior DLPFC. The data are sorted by the subjects' responses: utilitarian judgment (action judged "appropriate") = solid line; non-utilitarian judgment (action judged "inappropriate") = dashed line. Data are not adjusted for hemodynamic lag (Greene et al. 2004).

To summarize, people's moral judgments appear to be products of at least two different kinds of psychological processes. First, both brain imaging and reaction time data suggest that there are prepotent negative emotional responses that drive people to disapprove of the personally harmful actions proposed in cases like the *footbridge* and *crying baby* dilemmas. Second, further brain imaging results suggest that "cognitive" psychological processes can compete with the aforementioned emotional processes, driving people to approve of personally harmful moral violations, primarily when there is a strong utilitarian rationale for doing so, as in the *crying baby* case. The parts of the brain that exhibit increased activity when people make characteristically utilitarian judgments are those that are most closely associated with higher cognitive functions, such as executive control (Koechlin et al. 2003; Miller and Cohen 2001), complex planning (Koechlin et al. 1999), and deductive and inductive reasoning (Goel and Dolan 2004). These brain regions are among those most dramatically expanded in humans as compared to other primates (Allman et al. 2002).

This sort of "dual-process" model of decision-making, while new to moral psychology, has proven successful in other domains (Chaiken and Trope 1999;

Kahneman 2003; Lieberman et al. 2002). A common theme among these models is the idea that an individual's behavior is determined through an interaction between two processing streams operating in parallel: a quick and efficient processing stream that provides stereotyped responses based on limited information and a slower, more deliberative processing stream that provides more flexible responses based on a (potentially) much wider range of information. Recent neuroimaging studies have generated evidence for dual-process accounts of decision-making in the Ultimatum Game (Sanfey et al. 2003), decisions involving choices between immediate and delayed rewards (McClure et al. 2004), and the regulation of emotional responses to faces of members of racial out-groups (Cunningham et al. 2004). In each of these studies, several of the brain regions described above appear to perform functions similar to those described above.

Emotion and "Cognition" in Real-World Moral Judgment

In his classic article, "Famine, Affluence, and Morality," Peter Singer (1972) argues that we in the affluent world have an obligation to do much more than we do to improve the lives of needy people. He argues that if we can prevent something very bad from happening without incurring a comparable moral cost, then we ought to do it. For example, if someone notices a small child drowning in a shallow pond, he is morally obliged to wade in and save that child, even if it means muddying up his expensive clothes. As Singer points out, this seemingly innocuous principle has radical implications, implying that all of us who spend money on unnecessary luxuries should give up at least some of those luxuries in order to spend the money on saving and/or improving the lives of impoverished peoples. Why, Singer asks, do we have a strict obligation to save a nearby drowning child but no comparable obligation to save faraway sick and starving children through charitable donations to organizations like Oxfam?

Many normative explanations come to mind, but none is terribly compelling. Are we allowed to ignore the plight of faraway children because they are citizens of foreign nations? If yes, then would it be acceptable to let the child drown, provided that the child was encountered while traveling abroad, or in international waters? Is it because of diffused responsibility – because many are in a position to help a starving child abroad, but only you are in a position to help this hypothetical drowning child? What if there were many people standing around the pond doing nothing? Would that make it okay for you to do nothing as well? Is it because international aid is ultimately ineffective, only serving to enrich corrupt politicians, or create more poor people? In that case, our obligation would simply shift to more sophisticated relief efforts incorporating political reform, economic development, family planning education, and so on. Are all relief efforts doomed to ineffectiveness? That's a bold empirical claim that no one can honestly make with great confidence.

In response to Singer's moral challenge, we find ourselves in a similar position to the one we faced with the trolley problem. We have a strong intuition that two moral dilemmas are importantly different, and yet we have a hard time explaining what the crucial difference is (Kagan 1989; Unger 1996).

I believe that the same psychological theory that makes sense of the trolley problem can make sense of Singer's problem. Note that the interaction in the case of the drowning child is "up close and personal," the sort of situation that might have been encountered by our human and primate ancestors. Likewise, note that the donation case is not "up close and personal" and is not the sort of situation that our ancestors could have encountered. At no point were our ancestors able to save the lives of anonymous strangers through modest material sacrifices. In light of this, the psychological theory presented above suggests that we are likely to find the obligation to save the drowning child more pressing simply because this "up close and personal" case pushes our emotional buttons in a way that the more impersonal donation case does not (Greene 2003). As it happens, these two cases were among those tested in the brain imaging study described above, with a variation on the drowning child case included in the *personal* condition and a variation on the donation case included in the *impersonal* condition (Greene et al.2001, 2004.

Few people accept Singer's extreme utilitarian conclusion. Rather, people tend to believe that they are perfectly justified in spending their money on luxuries for themselves, despite the fact that their money could be used to dramatically improve the lives of other people. This is exactly what one would expect 1) if the sense of obligation is driven primarily by emotion, and 2) if, when it comes to obligations to aid, emotions are only sufficiently engaged when those to whom we might owe something are encountered (or conceived of) in a personal way.

However, the research presented above suggests that intuitive emotional responses need not dominate our moral thinking. It seems likely, for example, that Peter Singer has overridden his emotion-based judgments and replaced them, to some extent at least, with a kind of moral thinking that is more flexible and systematic. Moreover, it seems likely that the world would be better off if more people took Singer's argument seriously. (Singer currently gives about 20% of his annual income to charity.) Perhaps, a deeper scientific understanding of our ordinary moral thinking – where it comes from and the nature of the various forces that have shaped it – will incline us toward a kind of moral thinking that is, like Peter Singer's, extraordinary.

References

Adolphs, R (2003) Cognitive neuroscience of human social behaviour. Nat Rev Neurosci 4: 165–178.

Adolphs R, Tranel D, Damasio AR (1998) The human amygdala in social judgment. Nature 393: 470–474.

Allison T, Puce A, McCarthy G (2000) Social perception from visual cues: role of the STS region. Trends Cogn Sci 4: 267–278.

Allman J, Hakeem A, Watson K (2002) Two phylogenetic specializations in the human brain. Neuroscientist 8: 335–346.

Anderson SW, Bechara A, Damasio H, Tranel D, Damasio A R (1999) Impairment of social and moral behavior related to early damage in human prefrontal cortex. Nature Neurosci 2: 1032–1037.

Aquinas T (1988) Of killing. In: Baumgarth WP, Regan RJ (eds) On law, morality, and politics. Indianapolis/Cambridge: Hackett Publishing Co., pp. 226–227.

Bargh JA, Chartrand T. (1999) The unbearable automaticity of being. Am Psychol 54:, 462–479.

Bechara A, Damasio H, Damasio AR (2000) Emotion, decision making and the orbitofrontal cortex. Cereb Cortex 10: 295–307.

Blair RJ. (2001) Neurocognitive models of aggression, the antisocial personality disorders, and psychopathy. J Neurol Neurosurg Psychiat 71: 727–731.

Botvinick MM, Braver TS, Barch DM, Carter CS, Cohen JD (2001) Conflict monitoring and cognitive control. Psychol Rev 108: 624–652.

Chaiken S, Trope Y (eds) (1999) Dual-process theories in social psychology. New York: Guilford Press.

Cunningham WA, Johnson MK, Raye CL, Chris Gatenby J, Gore JC, Banaji MR (2004) Separable neural components in the processing of black and white faces. Psychol Sci 15: 806–813.

Damasio AR (1994) Descartes' error: Emotion, reason, and the human brain. New York: G.P. Putnam.

Damasio A (2003) Looking for Spinoza: joy, sorrow, and the feeling brain. New York: Harcourt.

Fischer JM, Ravizza M (eds) (1992) Ethics: problems and principles. Fort Worth, TX: Harcourt Brace Jovanovich College Publishers).

Foot P (1978) The problem of abortion and the doctrine of double effect. In: Virtues and vices. Oxford: Blackwell.

Goel V, Dolan RJ (2004) Differential involvement of left prefrontal cortex in inductive and deductive reasoning. Cognition 93: B109–121.

Greene J. (2003) From neural 'is' to moral 'ought': what are the moral implications of neuroscientific moral psychology? Nature Rev Neurosci 4: 846–849.

Greene JD, Sommerville R., Nystrom LE, Darley JM, Cohen JD (2001) An fMRI investigation of emotional engagement in moral judgment. Science 293: 2105–2108.

Greene JD, Nystrom LE, Engell AD, Darley JM, Cohen JD (2004) The neural bases of cognitive conflict and control in moral judgment. Neuron 44: 389–400.

Haidt J (2001) The emotional dog and its rational tail: A social intuitionist approach to moral judgment. Psychol Rev 108: 814–834.

Kagan S (1989) The limits of morality. New York, Oxford University Press.

Kahneman D (2003) A perspective on judgment and choice: mapping bounded rationality. Am Psychol 58: 697–720.

Kant I (1959) Foundation of the metaphysics of morals. Indianapolis: Bobbs-Merrill).

Kiehl KA, Smith AM, Hare RD, Mendrek A, Forster BB, Brink J, Liddle PF (2001) Limbic abnormalities in affective processing by criminal psychopaths as revealed by functional magnetic resonance imaging. Biol Psychiat 50: 677–684.

Koechlin E, Basso G, Pietrini P, Panzer S, Grafman J (1999) The role of the anterior prefrontal cortex in human cognition. Nature 399: 148–151.

Koechlin E, Ody C, Kouneiher F (2003) The architecture of cognitive control in the human prefrontal cortex. Science 302: 1181–1185.

Kohlberg L (1969) Stage and sequence: the cognitive-developmental approach to socialization. In: Goslin DA (ed) Handbook of socialization theory and research. Chicago: Rand McNally, pp. 347–480.

Lieberman MD, Gaunt R, Gilbert DT, Trope Y (2002) Reflection and reflexion: a social cognitive neuroscience approach to attributional inference. Advan Exp Soc Psychol 34: 199–249.

Maddock RJ (1999) The retrosplenial cortex and emotion: new insights from functional neuroimaging of the human brain. Trends Neurosci 22: 310–316.

McClure SM, Laibson DI, Loewenstein G, Cohen JD (2004) Separate neural systems value immediate and delayed monetary rewards. Science 306: 503–507.

Mill JS (1998) Utilitarianism. In: Crisp R (ed) New York: Oxford University Press.

Miller EK, Cohen JD (2001) An integrative theory of prefrontal cortex function. Annu Rev Neurosci 24: 167–202.

Moll J, Eslinger PJ, Oliveira-Souza R (2001) Frontopolar and anterior temporal cortex activation in a moral judgment task: preliminary functional mri results in normal subjects. Arq Neuropsiquiatr 59: 657–664.

Moll J, de Oliveira-Souza R, Bramati I, Grafman J (2002a) Functional networks in emotional moral and nonmoral social judgments. Neuroimage 16: 696.

Moll J, de Oliveira-Souza R, Eslinger PJ, Bramati IE, Mourao-Miranda J, Andreiuolo PA, Pessoa L (2002b) The neural correlates of moral sensitivity: a functional magnetic resonance imaging investigation of basic and moral emotions. J Neurosci 22: 2730–2736.

Petrinovich L, O'Neill P (1996) Influence of wording and framing effects on moral intuitions. Ethol Sociobiol 17: 145–171.

Petrinovich L, O'Neill P, Jorgensen M (1993) An empirical study of moral intuitions: toward an evolutionary ethics. J Pers Soc Psychol 64: 467–478.

Phan KL, Wager T, Taylor SF, Liberzon I. (2002) Functional neuroanatomy of emotion: A meta-analysis of emotion activation studies in PET and fMRI. Neuroimage 16: 331–348.

Rozin P, Lowery L, Imada S, Haidt J (1999) The cad triad hypothesis: a mapping between three moral emotions (contempt, anger, disgust) and three moral codes (community, autonomy, divinity). J Pers Soc Psychol 76: 574–586.

Sanfey AG, Rilling JK, Aronson JA, Nystrom LE, Cohen JD (2003) The neural basis of economic decision-making in the ultimatum game. Science 300: 1755–1758.

Saxe R, Carey S, Kanwisher N (2004) Understanding other minds: liking developmental psychology and functional neuroimaging. Ann Rev Psychol 55: 87–124.

Singer P (1972) Famine, affluence and morality. Philos Publ Affairs 1: 229–243.

Sober E, Wilson DS (1998) Unto others: the evolution and psychology of unselfish behavior. Cambridge, Mass: Harvard University Press.

Thomson JJ (1986) Rights, restitution, and risk: essays, in moral theory. Cambridge, Mass.: Harvard University Press.

Trivers RL (1971) The evolution of reciprocal altruism. Quart Rev Biol 46: 35–57.

Unger PK (1996) Living high and letting die: our illusion of innocence. New York, NY: Oxford University Press.

Wrangham R, Peterson D (1996) Demonic males: apes and the origins of human violence. Boston: Houghton Mifflin.

Neural substrates of affective style and value

Richard J. Davidson[1]

Introduction

One of the most salient features of emotion is the pronounced variability among individuals in their reactions to emotional incentives and in their dispositional mood. Collectively, these individual differences have been described as affective style. Some types of affective style promote more empathic and compassionate behavior than others. Recent research has begun to dissect the constituents of affective style. The search for these components is guided by the neural systems that instantiate emotion and emotion regulation. In this essay, this body of research and theory is applied specifically to positive affect and well-being. The central substrates and peripheral biological correlates of well-being are described. A resilient affective style is associated with high levels of left prefrontal activation, effective modulation of activation in the amygdala and fast recovery in response to negative and stressful events. In peripheral biology, these central patterns are associated with lower levels of basal cortisol and with higher levels of antibody titers to influenza vaccine. The essay concludes with a consideration of whether these patterns of central and peripheral biology can be modified by training and shifted toward a more salubrious direction.

One of the most salient characteristics of emotion is the remarkable heterogeneity among individuals in how they respond to the same emotionally provocative challenge. Such differences in patterns of emotional reactivity play a crucial role in shaping variations in well-being. While individual differences in emotion processing can be found at many levels of phylogeny, they are particularly pronounced in primates and probably are most extreme in humans. A number of evolutionary theorists have speculated on the adaptive significance of such individual differences (Wilson 1994). While these arguments have never been applied to the domain of emotion and affective style, it is not difficult to develop hypotheses about how such differences might provide advantages to individuals living in groups. Though these distal influences are interesting and important, there is

[1] Laboratory for Affective Neuroscience, University of Wisconsin, 1202 West Johnson Street, Madison, WI 53706, USA; e-mail: rjdavids@wisc.edu

Changeux et al.
Neurobiology of Human Values
© Springer-Verlag Berlin Heidelberg 2005

a paucity of empirical findings that bear on these issues. This essay will mostly focus on the proximal mechanisms that underlie such individual differences, with a focus on well-being. The central substrates of individual differences in components of well-being will be described. The possible influence of the central circuitry of emotion on peripheral biological indices that are relevant to physical health and illness will also be considered. Finally, the concluding section acknowledges the important advances that have recently been made in our understanding of neuroplasticity, and in this section I argue that it would be best to conceptualize emotional characteristics such as happiness as skills that can be trained in ways that are not fundamentally different from other kinds of skill learning.

Affective style refers to consistent individual differences in emotional reactivity and regulation (see Davidson 1998a; Davidson et al. 2000a,b). It is a phrase that is meant to capture a broad array of processes that either singly or in combination modulate an individual's response to emotional challenges, dispositional mood and affect-relevant cognitive processes. Affective style can refer to valence-specific features of emotional reactivity or mood or it can refer to discrete emotion-specific features. Both levels of analysis are equally valid and the choice of level should be dictated by the question posed.

Rapid developments in our understanding of emotion, mood and affective style have come from the study of the neural substrates of these phenomena. The identification of the brain circuitry responsible for different aspects of affective processing has helped to parse the domain of emotion into more elementary constituents in a manner similar to that found in cognitive neuroscience, where an appeal to the brain has facilitated the rapid development of theory and data on the subcomponents of various cognitive processes (see, e.g., Kosslyn and Koenig 1992).

Both lesion and neuroimaging studies provide information primarily about the "where" question, that is, where in the brain are computations related to specific aspects of affective processing occurring? It is important at the outset to consider both the utility of knowing "where" and how such information can provide insight into the "how" question, that is, how might a particular part of the brain instantiate a specific process that is essential to affective style? The brain sciences are now replete with information about the essential nature of specific types of information processing in different regions of the brain. For example, there is evidence to suggest that the dorsolateral prefrontal cortex (DLPFC) is important for maintaining a representation of information on-line in the absence of immediate cues. The neurophysiological basis of this type of information processing has been actively studied in the animal laboratory (e.g., Goldman-Rakic 1996, 2000). If this region of the brain is activated at certain times in the stream of affective information processing, we can develop hypotheses on the basis of extant work about what this territory of the prefrontal cortex (PFC) might be doing during the affective behavior and how it might be doing it. A related consideration is the network of anatomical connectivity to and from a particular brain region. From a consideration of connectivity, insights may be gleaned as to how a particular brain region might react during a particular form of emotional processing. For example, we know that regions of the amygdala have extensive connectivity with cortical territories that can become activated following activation of the amygdala. In this way, the amygdala can issue a cortical call for further processing in response to potentially threatening stimuli that must be processed further to assess danger. There are prefrontal regions that have extensive anatomical connectivity with the amygdala and they appear to play a modulatory role over amygdala function. Other regions of the amygdala have extensive connections to limbic and brain stem circuits that can modulate behavioral and autonomic outflow. Adjustments in autonomic responses and action tendencies are typical components of emotion.

Conceptual and methodological considerations in the study of affective style

Current research on well-being is largely based on the use of self-report measures to make inferences about variation among individuals in type and magnitude of well-being. One important component of neurobiological research on well-being is to begin to dissect well-being into more specific constituents that may underlie the coarse phenomenological descriptions provided by subjects. In addition, research on the neural correlates of well-being may provide an independent biological measure sensitive to variations in well-being that are not subject to the kinds of reporting and judgmental biases commonly found in the self-report measures. For example, researchers have found that questions that precede items asking about well-being can influence a subject's report of well-being. Variations in the weather can similarly affect such reports. These examples illustrate the fact that, when subjects are queried about global well-being, they frequently utilize convenient heuristics to answer such questions and typically do not engage in a systematic integration of utility values over time. It may be that certain parameters of brain function are better repositories of the cumulative experiences that inevitably shape well-being. At the present point in the development of this science, these are mere speculations in search of evidence, but the time is ripe for such evidence to be gathered.

The status of research on well-being is now at a point occupied about a decade ago or more by research on mood and anxiety disorders, though it continues to suffer from some of the same problems. Mood and anxiety disorders are generally conceptualized as being caused or at least accompanied by dysfunctions of emotion. However, what specific affective process is dysfunctional is rarely, if ever, delineated, and nosological schemes for categorizing these disorders do not rely upon the specific nature of the affective dysfunction in question but rather are based upon phenomenological description. Research in my laboratory over the past 15 years has been predicated on the view that more meaningful and rapid progress in understanding the brain bases of mood and anxiety disorders can be achieved if we move to an intermediate level of description that penetrates below the categorical, phenomenologically based classifications of the DSM and seeks to characterize the specific nature of the affective styles that are associated with vulnerability to these forms of psychopathology.

Many of the parameters of affective style, such as the threshold to respond, magnitude of response, latency to peak of response, and recovery function, are features that are often opaque to conscious report, though they may influence the subjective experience of emotion. These parameters of responding can be measured in many different response systems, including both central and peripheral systems. For example, magnitude of response can be measured in a peripheral measure, such as the emotion-modulated startle (Lang 1995), or in a central measure, such as activation in the amygdala assessed with functional magnetic resonance imaging (fMRI). The extent to which coherence across response systems in these parameters is present has not yet been systematically addressed. In previous work, we have argued that variations in some of these parameters in particular response systems are especially relevant to vulnerability to mood, anxiety and other disorders and also to resilience (e.g., Davidson et al. 2000a, b). One of the important developments in emotion research in general and in affective neuroscience in particular is the capacity to objectively measure these parameters of responding. For example, in several studies, we have used the emotion-modulated startle to capture the time course of valence-specific emotion responding (Jackson et al. 2000a; Larson et al. 1998). The startle reflex is controlled by a brainstem circuit that is influenced by activity in forebrain structures. Davis (1992) elegantly dissected the circuitry through which the magnitude of this reflex is modulated during the arousal of fear in rodents, demonstrating that it is via a descending pathway from the central nucleus of the amygdala to the nucleus pontine reticularis in the brain stem that the magnitude of startle is enhanced in response to a conditioned fear cue. Lesions of the central nucleus of the amygdala abolish the fear potentiation of the startle but do not affect the magnitude of the baseline startle. Lang and his colleagues (Vrana et al. 1988) were the first to systematically show that the same basic phenomenon can be produced in humans. They took advantage of the fact that brief acoustic noise bursts produce the eyeblink component of the startle and little else, thus enabling their presentation as innocuous stimuli in the background. By measuring electromyographic activity from the orbicularis oculii muscle with two miniature electrodes under one eye, they were able to quantify the strength of the blink response and show that the magnitude of the blink was greater when subjects were presented with unpleasant pictures in the foreground, compared with the presentation of neutral pictures. Moreover, when subjects were exposed to positive stimuli, the magnitude of startle was actually attenuated relative to a neutral condition (Vrana et al. 1988). This same basic effect has now been reported with many different types of foreground stimuli in several modalities (see Lang 1995, for review).

We have exploited the emotion-modulated startle to begin to characterize the time course of affective responding, or what I have referred to as affective chronometry (Davidson 1998a). By inserting acoustic noise probes at different latencies before and after a critical emotional stimulus is presented, both the anticipatory limb as well as the recovery limb of the response can be measured. And, by utilizing paradigms in the MRI scanner that were first studied in the psychophysiology laboratory, the neural circuitry underlying the different phases of affective processing can be interrogated with fMRI. Our current work in this area has emphasized the importance of the recovery function following negative events for vulnerability to certain forms of psychopathology as well as for resilience. We have argued that the failure to rapidly recover following a negative event can be a crucial ingredient of vulnerability to both anxiety and mood disorders, particularly when such a style is combined with frequent exposure to negative events over a sustained period of time. The failure to adequately recover would result in sustained elevations in multiple systems that are activated in response to negative events. On the other hand, the capacity for rapid recovery following negative events may define an important ingredient of resilience. We have defined resilience as the maintenance of high levels of positive affect and well-being in the face of significant adversity. It is not that resilient individuals never experience negative affect, but rather that the negative affect does not persist. Such individuals are able to profit from the information provided by the negative affect, and their capacity for "meaning making" in response to such events may be part and parcel of their ability to show rapid decrements in various biological systems following exposure to a negative or stressful event (see, e.g., Geise-Davis and Spiegel 2003).

Neural substrates of emotion and affective style

In this section, a brief overview is provided of core components of the circuitry that instantiates some important aspects of emotion and affective style, with an emphasis on prefrontal cortex and the amygdala. It is not meant to be an exhaustive review but rather will present selected highlights to illustrate some of the key advances that have been made recently.

Emotion and affective style are governed by a circuit that includes the following structures and likely others as well: dorsolateral prefrontal cortex (DLPFC), ventromedial PFC (vmPFC), orbitofrontal cortex (OFC), amygdala, hippocampus, anterior cingulate cortex (ACC) and insular cortex. It is argued that different subprocesses are instantiated in each of these structures and that they normally work together to process, generate and regulate emotional information and emotional behavior.

Prefrontal cortex

A large corpus of data at both the animal and human levels implicates various sectors of the PFC in emotion. The PFC is not a homogeneous zone of tissue but rather has been differentiated on the basis of both cytoarchitectonic and functional considerations. The three subdivisions of the primate PFC that have been consistently distinguished are the DLPFC, vmPFC, and OFC. In addition, there appear to be important functional differences between the left and right sides within some of these sectors.

The case for the differential importance of left and right PFC sectors for emotional processing was first made systematically in a series of studies on patients with unilateral cortical damage (Gainotti 1972; Robinson et al. 1984; Sackeim et al. 1982). These findings, as well as more modern efforts to examine mood consequences of unilateral prefrontal damage, have been extensively reviewed by Davidson (2004) and thus will not be considered here.

A growing corpus of evidence in normal intact humans is consistent with the findings derived from the lesion evidence. Davidson and his colleagues have reported that induced positive and negative affective states shift the asymmetry in prefrontal brain electrical activity in lawful ways. For example, film-induced negative affect increases relative right-sided prefrontal and anterior temporal activation (Davidson et al. 1990), whereas induced positive affect elicits an opposite pattern of asymmetric activation. Similar findings have been obtained by others (e.g., Ahern and Schwartz 1985; Jones and Fox 1992).

Using a cued reaction time paradigm with monetary incentives, Sobotka et al. (1992) first reported that, in the anticipatory interval between the cue and the response, EEG differences were observed between reward and punishment trials, with greater left-sided frontal activation observed in response to the former compared with the latter trial type. In a more recent study, Miller and Tomarken (2001) replicated and extended this basic effect and very recently, we (Shackman et al. 2003) replicated the Miller and Tomarken effect, showing that reward trials produced significantly greater left prefrontal activation in the anticipatory interval compared with no incentive trials. In addition to these studies that manipulated phasic emotion, we will review in the next section a body of evidence that supports the conclusion that individual differences in baseline levels of asymmetric activation in these brain regions are lawfully related to variations in dispositional affective style. Using an extended picture presentation paradigm designed to evoke longer-duration changes in mood (Sutton et al. 1997a), we measured regional glucose metabolism with positron emission tomography (PET) to ascertain whether similar patterns of anterior asymmetry would be present using this very different and more precise method to assess regional brain activity (Sutton et al. 1997b). During the production of negative affect, we observed right-sided increases in metabolic rate in anterior orbital, inferior frontal, middle and superior frontal gyri, whereas the production of positive affect was associated with a pattern of predominantly left-sided metabolic increases in the pre- and postcentral gyri. Using PET to measure regional cerebral blood flow, Hugdahl and his colleagues (1995; Hugdahl 1998) reported a widespread zone of increased blood flow in the right PFC, including the orbitofrontal and dorsolateral cortices and inferior and superior cortices, during the extinction phase, after aversive learning had occurred, compared with the habituation phase, prior to the presentation of the experimental contingencies.

Other investigators have used clinical groups to induce a stronger form of negative affect in the laboratory than is possible with normal controls. One common strategy for evoking anxiety among anxious patients in the laboratory is to present them with specific types of stimuli that are known to provoke their anxiety (e.g., pictures of spiders for spider phobics; making a public speech for social phobics). Davidson and colleagues (2000d), in a study using brain electrical activity measures, have recently found that when social phobics anticipate making a public speech, they show large increases in right-sided anterior activation. Pooling across data from three separate anxiety-disordered groups that were studied with PET, Rauch and his colleagues (1997) found two regions of the PFC that were consistently activated across groups: the right inferior PFC and right medial orbital PFC.

The ventromedial PFC has been implicated in the anticipation of future positive and negative affective consequences. Bechara and his colleagues (1994) reported that patients with bilateral lesions of the ventromedial PFC have difficulty anticipating future positive or negative consequences, although immediately available rewards and punishments do influence their behavior. Such patients show decreased levels of electrodermal activity in anticipation of a risky choice compared with controls, whereas controls exhibit such an autonomic change before they explicitly know that it is a risky choice (Bechara et al. 1996, 1997, 1999).

The findings from the lesion method – when the effects of small unilateral lesions are examined – and from neuroimaging studies in normal subjects and patients with anxiety disorders converge on the conclusion that increases in right-sided activation in various sectors of the PFC are associated with increased negative affect. Less evidence is available for the domain of positive affect, in part because positive affect is much harder to elicit in the laboratory and because of the negativity bias (see Cacioppo and Gardner 1999; Taylor 1991). This latter phenomenon refers to the general tendency of organisms to react more strongly to negative compared to positive stimuli, perhaps as a consequence of evolutionary pressures to avoid harm.

Systematic studies designed to disentangle the specific role played by various sectors of the PFC in emotion are lacking. Many theoretical accounts of emotion assign it an important role in guiding action and organizing behavior toward the acquisition of motivationally significant goals (e.g.,Frijda 1994; Levenson 1994). This process requires that the organism have some means of representing affect in the absence of immediately present rewards and punishments and other affective incentives. It is likely that the PFC plays a key role in this process (see Watanabe 1996). Damage to certain sectors of the PFC impairs an individual's capacity to anticipate future affective outcomes and consequently results in an inability to guide behavior in an adaptive fashion. Such damage is not likely to disrupt an individual's responding to immediate cues for reward and punishment, only the anticipation before and maintenance after an affective cue is presented. This proposal can be tested using current neuroimaging methods (e.g., fMRI) but has not yet been rigorously evaluated. With regard to the different functional roles of the dorsolateral, orbitofrontal and ventromedial sectors of the PFC, Davidson and Irwin (1999) suggested, on the basis of considering both human and animal studies, that the ventromedial sector is most likely involved in the representation of elementary positive and negative affective states in the absence of immediately present incentives. The ventromedial sector also has strong anatomical reciprocity with the amygdala and thus appears to play an important role in modulating activity in the amygdala, thereby contributing to the regulation of emotion. The orbitofrontal sector has most firmly been linked to rapid learning and unlearning of stimulus-incentive associations and has been particularly implicated in reversal learning (Rolls 1999). Therefore, the orbitofrontal sector is also likely key to understanding aspects of emotion regulation (see Davidson et al. 2000b). One critical component of emotion regulation is the relearning of stimulus-incentive associations that might have been previously maladaptive, a process likely requiring the orbitofrontal cortex. The dorsolateral sector is most directly involved in the representation of goal states toward which more elementary positive and negative states are directed.

Amygdala

A large corpus of research at both the animal and human levels has established the importance of the amygdala for emotional processes (LeDoux 1996; Cahill and McGaugh 1998; Aggleton 1993; Davis and Whalen 2001). Since many reviews of the animal literature have appeared recently, a detailed description of these studies will not be presented here. LeDoux and his colleagues marshaled a large corpus of compelling evidence to suggest that the amygdala is necessary for the establishment of conditioned fear. Whether the amygdala is necessary for the expression of that fear following learning and whether the amygdala is the actual locus of where the learned information is stored are still matters of some controversy (see Cahill et al. 1999; Fanselow and LeDoux 1999). The classic view of amygdala damage in non-human primates resulting in major affective disturbances – as expressed in the Kluver-Bucy syndrome, where the animal exhibits abnormal approach, hyper-orality and sexuality, and little fear – is now thought to be a function of damage elsewhere in the medial temporal lobe. When very selective excitotoxic lesions of the amygdala are made that preserve fibers of passage, nothing resembling the Kluver-Bucy syndrome is observed (Kalin et al. 2001). The upshot of this diverse array of findings is to suggest a more limited role for the amygdala in certain forms of emotional learning, though the human data imply a more heterogeneous contribution.

While the number of patients with discrete lesions of the amygdala is small, they have provided unique information on the role of this structure in emotional processing. Several studies have now reported specific impairments in the recognition of facial expressions of fear in patients with restricted amygdala damage (Adolphs et al. 1995; 1996; Broks et al. 1998; Calder et al. 1996). Interestingly, in a very recent report, Adolphs and his colleagues (2005) specifically identified a failure to gaze normally at the eye region of the face as the proximal cause of the deficit in fear face recognition in patient SM, a patient with bilateral destruction of the amygdala. Recognition of facial signs of other emotions has been found to be mostly intact. In a study that required subjects to make judgments about the trustworthiness and approachability of unfamiliar adults from facial photographs, patients with bilateral amygdala damage judged the unfamiliar individuals to be more approachable and trustworthy than did control subjects (Adolphs et al. 1998). Recognition of vocal signs of fear and anger was found to be impaired in a patient with bilateral amygdala damage (Scott et al. 1997), suggesting that this deficit is not restricted to facial expressions. Other researchers (Bechara et al. 1995) have demonstrated that aversive autonomic conditioning is impaired in a patient with amygdala damage despite the fact that the patient showed normal declarative knowledge of the conditioning contingencies. Collectively, these findings from patients with selective bilateral destruction of the amygdala suggest specific impairments on tasks that tap aspects of negative emotion processing. Most of the studies have focused on the perceptual side, where the data clearly show the amygdala to be important for the recognition of cues of threat or danger. The conditioning data also indicate that the amygdala may be necessary for acquiring new implicit autonomic learning of stimulus-punishment contingencies. In one of the few studies to examine the role of the amygdala in the expression of already learned emotional responses, Angrilli and colleagues (1996) reported on a patient with a benign tumor of the right amygdala in a study that used startle magnitude in response to an acoustic probe measured from orbicularis oculi. Among control subjects, they observed the well-known effect of startle potentiation during the presentation of aversive stimuli. In the patient with right amygdala damage, no startle potentiation was observed in response to aversive versus neutral stimuli. These findings suggest that the amygdala might be necessary for the expression of already learned negative affect.

Since 1995, a growing number of studies using PET and fMRI to investigate the role of the amygdala in emotional processes have begun to appear. Many studies have reported activation of the amygdala detected with either PET or fMRI when anxiety-disordered patients have been exposed to their specific anxiety-provoking stimuli compared with control stimuli (e.g., Breiter et al. 1996b; Rauch et al. 1996). When social phobics were exposed to neutral faces, they showed activation of the amygdala comparable to what was observed in both the phobics and controls in response to aversive compared with neutral odors (Birbaumer et al. 1998). Consistent with the human lesion data, a number of studies have now reported activation of the amygdala in response to facial expressions of fear compared with neutral, happy or disgust control faces (Morris et al. 1996; Phillips et al. 1997). In the fMRI study by Brieter et al. (1996 a), they observed rapid habituation of the amygdala response, which may provide an important clue to the time-limited function of the amygdala in the stream of affective information processing. Whalen and his colleagues (1998) observed activation of the amygdala in response to masked fear faces that were not consciously perceived. Unpleasant compared with neutral and pleasant pictures have also been found to activate the amygdala (Irwin et al. 1996). Finally, a number of studies have reported activation of the amygdala during early phases of aversive conditioning (Buchel et al. 1998; LaBar et al. 1998). Amygdala activation in response to several other experimental procedures for inducing negative affect has been reported, including unsolvable anagrams of the sort used to induce learned helplessness (Schneider et al. 1996), aversive olfactory cues (Zald and Pardo 1997) and aversive gustatory stimuli (Zald et al. 1998). Other data on individual differences in amygdala activation and their relation to affective style will be treated in the next section. The issues of whether the amygdala responds preferentially to aversive versus appetitive stimuli, is functionally asymmetric, and is required for both the initial learning and subsequent expression of negative emotional associations have not yet been adequately resolved and are considered in detail elsewhere (Davidson and Irwin 1999), though some data clearly suggest that the amygdala does activate in response to appetitive stimuli (Hamman et al. 2002). It should be noted that one fMRI study (Zalla et al. 2000) found differential activation of the left and right amygdala to winning and losing money, with the left amygdala showing increased activation to winning more money whereas the right amygdala showed increased activation in response to the parametric manipulation of losing money. Systematic examination of asymmetries in amygdala activation and function in appetitive and aversive contexts should be performed in light of these data. In several recent reviews, Whalen (Davis and Whalen 2001) has argued that a major function of the amygdala is the detection of ambiguity and the issuing of a call for further processing when ambiguous information is presented. I will return to this claim later in this chapter, when the issue of individual differences is addressed.

These findings raise the question concerning the "optimal" pattern of amygdala function for well-being. Based upon evidence reviewed below in the context of individual differences, we will argue that low basal levels of amygdala activation in conjunction with situationally appropriate responding, effective top-down regulation and rapid recovery characterize a pattern that is consistent with high levels of well-being.

What are individual differences in PFC and amygdala activations associated with?

In both infants (Davidson and Fox 1989) and adults (Davidson and Tomarken 1989), there are large individual differences in baseline electrophysiological measures of prefrontal activation, and such individual variation is associated with differences in aspects of affective reactivity. In infants, Davidson and Fox (1989) reported that 10-month-old babies who cried in response to maternal separation were more likely to have less left- and greater right-sided prefrontal activation during a preceding resting baseline compared with infants who did not cry in response to this challenge. In adults, we first noted that the phasic influence of positive and negative emotion elicitors (e.g., film clips) on measures of prefrontal activation asymmetry appeared to be superimposed upon more tonic individual differences in the direction and absolute magnitude of asymmetry (Davidson and Tomarken 1989).

During our initial explorations of this phenomenon, we needed to determine if baseline electrophysiological measures of prefrontal asymmetry were reliable and stable over time and thus could be used as a trait-like measure. Tomarken et al. (1992) recorded baseline brain electrical activity from 90 normal subjects on two occasions separated by approximately three weeks and found excellent internal consistency reliability and adequate test-retest stability for metrics of prefrontal activation asymmetry over this time period.

On the basis of our prior data and theory, we reasoned that extreme left- and extreme right-frontally activated subjects would show systematic differences in dispositional positive and negative affect. We administered the trait version of the Positive and Negative Affect Scales (PANAS; Watson et al. 1988) to examine this question and found that the left-frontally activated subjects reported more positive and less negative affect than their right-frontally activated counterparts (Tomarken et al. 1992; see Fig. 1). More recently (Sutton and Davidson 1997), we showed that scores on a self-report measure designed to operationalize Gray's concepts of Behavioral Inhibition and Behavioral Activation (the BIS/BAS scales; Carver and White 1994) were even more strongly predicted by electrophysiological measures of prefrontal asymmetry than were scores on the PANAS scales. Subjects with greater left-sided prefrontal activation reported more relative BAS to BIS activity compared with subjects exhibiting more right-sided prefrontal activation.

In a very recent study, we extended these early findings and found that baseline measures of asymmetric prefrontal activation predicted reports of well-being

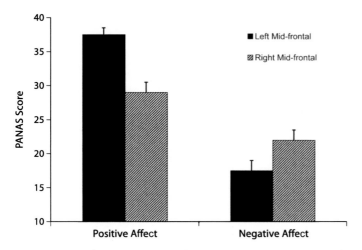

Fig. 1. Dispositional positive affect (from scores on the PANAS-General Positive Affect Scale) in subjects who were classified as extreme and stable left-frontally active (N=14) and extreme and stable right-frontally active (N=13) on the basis of electrophysiological measures of baseline activation asymmetries on two occasions separated by three weeks. Error bars denote standard error of the mean (Tomarken et al. 1992).

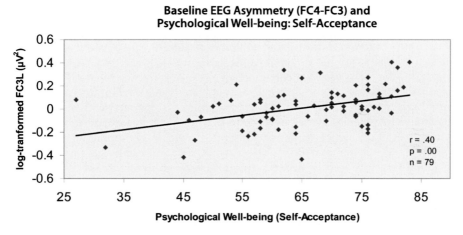

Fig. 2. Scatterplot depicting the correlation between frontal EEG asymmetry (FC4 - FC3) and total psychological well-being. Relative left frontal asymmetry (denoted by positive values on the abscissa) is associated with higher levels of well-being. (Urry et al. 2004).

among individuals in their late 50s (Urry et al. 2004; see Fig. 2). Moreover, this association was present even when the association between prefrontal activation asymmetry and dispositional positive affect was statistically removed. These findings indicate that prefrontal activation asymmetry accounts for variance in well-being over and above that which is accounted for by positive affect.

In addition to the studies described above using self-report and psychophysiological measures of emotion, we have also examined relations between individual differences in electrophysiological measures of prefrontal asymmetry and other biological indices that in turn have been related to differential reactivity to stressful events. Three recent examples from our laboratory include measures of immune function, cortisol and corticotropin-releasing hormone (CRH). The latter two measures represent key molecules in the activation of a coordinated response to stressful events. Our strategy in each case was to examine relations between individual differences in measures of prefrontal activation asymmetry and these other biological indices. In two separate studies (Kang et al. 1991; Davidson et al. 1999), we examined relations between the prefrontal activation indices and natural killer (NK) activity, since declines in NK activity have been reported in response to stressful, negative events (Kiecolt-Glaser and Glaser 1981). We predicted that subjects with greater left-sided prefrontal activation would exhibit higher NK activity compared with their right-activated counterparts, because the former type of subject has been found to report more dispositional positive affect, to show higher relative BAS activity and to respond more intensely to positive emotional stimuli. In each of the two studies conducted with independent samples, we found that left-frontally activated subjects indeed had higher levels of NK activity compared to their right-frontally activated counterparts (Kang et al. 1991; Davidson et al. 1999). We also examined the magnitude of change in NK activity in response to stress and found that subjects with greater baseline levels of left prefrontal activation showed the smallest magnitude decline in NK activity in response to stress compared with other subjects (Davidson et al. 1999).

One of the concerns with the studies that examine NK function is the fact that this is an in vitro assay and its significance for immunocompetence is unclear. To address this concern, we recently completed a study examining relations between prefrontal activation asymmetry and antibody responses to influenza vaccine (Rosenkranz et al. 2003) in a middle-aged sample of 52 subjects with an average age of 58 years (evenly divided by sex). In this study, we recorded brain electrical measures in the same way as previously described. We compared individuals in the top and bottom quartile on measures of prefrontal activation asymmetry and found large differences between these extreme groups in antibody titers to influenza vaccine (see Fig. 3), with the left-prefrontally activated subjects showing significantly greater antibody titers compared with their right-prefrontally activated counterparts.

In collaboration with Kalin, our laboratory has been studying similar individual differences in scalp-recorded measures of prefrontal activation asymmetry in rhesus monkeys (Davidson et al. 1992, 1993). We (Kalin et al. 1998) acquired measures of brain electrical activity from a large sample of rhesus monkeys (N=50). EEG measures were obtained during periods of manual restraint. A subsample of 15 of these monkeys was tested on two occasions four months apart. We found that the test-retest correlation for measures of prefrontal asymmetry was 0.62, suggesting similar stability of this metric in monkey and man. In the group of 50 animals, we also obtained measures of plasma cortisol during the early morning. We hypothesized that if individual differences in prefrontal asymmetry were associated with dispositional affective style, such differences should be correlated with cortisol, since individual differences in baseline cortisol have been related to various aspects of trait-related stressful behavior and psychopathology (see, e.g., Gold et al. 1988). We found that animals with left-sided prefrontal activation had lower levels of baseline cortisol than their right-frontally activated counterparts (see Fig. 4). As can be seen from the figure, it is the left-activated animals that are particularly low compared with both middle and right-activated subjects. More-

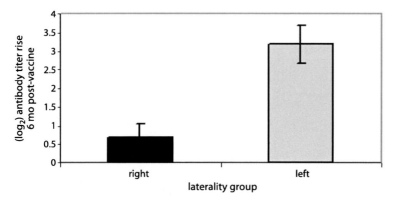

Fig. 3. Bar graph of the mean antibody titer rise (log$_2$) to influenza vaccine six months post-vaccine for extreme groups comprised of individuals (average age of 58 years) in the top and bottom 25th percentiles of activation asymmetry at the lateral frontal (F7/8) site. Error bars denote standard error of the mean. The difference between groups was highly significant ($t(22)$ =3.81, p<.001). (Rosenkranz et al. 2003).

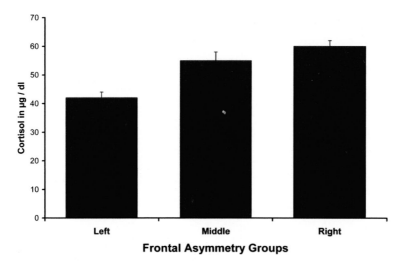

Fig. 4. Basal morning plasma cortisol from one-year-old rhesus monkeys classified as left- (N=12), middle- (N=16) or right- (N=11) frontally activated based upon electrophysiological measurements. Error bars denote standard error of the mean. (Kalin et al.1998).

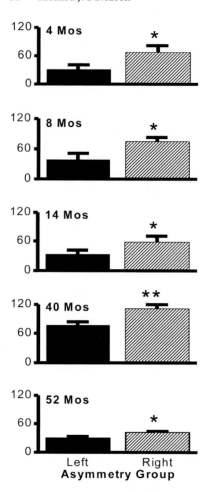

Fig. 5. Differences between right- (N=9) and left-prefrontally (N=10) activated animals in cerebrospinal fluid measures of corticotropin releasing hormone at five different ages. Units are in pg/ml and error bars denote standard error of the mean. The original classification of the animals as extreme right or left activated was performed on the basis of brain electrical activity data collected when the animals were 13 months of age. (Kalin et al. 2000).

over, when blood samples were collected two years following our initial testing, animals classified as showing extreme left-sided prefrontal activation at age one year had significantly lower baseline cortisol levels when they were three years of age compared with animals who were classified at age one year as displaying extreme right-sided prefrontal activation. Similar findings were obtained with cerebrospinal fluid levels of CRH. Those animals with greater left-sided prefrontal activation showed lower levels of CRH (Kalin et al. 2000; see Fig. 5). These findings indicate that individual differences in prefrontal asymmetry are present in non-human primates and that such differences predict biological measures that are related to affective style.

With the advent of neuroimaging, it has become possible to investigate the relation between individual differences in aspects of amygdala function and measures of affective style. We have used PET with flourodeoxyglucose (FDG) as a tracer to investigate relations between individual differences in glucose metabolism in the amygdala and dispositional negative affect. FDG-PET is well-suited to capture trait-like effects, since the period of active uptake of tracer in the brain is approximately 30 minutes. Thus, it is inherently more reliable than O^{15} blood flow measures, since the FDG data reflect activity aggregated over a 30-minute period. We have used resting FDG-PET to examine individual differences in glucose metabolic rate in the amygdala and its relation to dispositional negative affect in depressed subjects (Abercrombie et al.1998). We acquired a resting FDG PET scan as well as a structural MR scan for each subject. The structural MR scans are used for anatomical localization by coregistering the two image sets. Thus, for each subject, we used an automated algorithm to fit the MR scan to the PET image. Regions of interest (ROIs) were then drawn on each subject's MR scan to outline the amygdala in each hemisphere. These ROIs were drawn on coronal sections of subjects' MR images and the ROIs were then automatically transferred to the co-registered PET images. Glucose metabolism in the left and right amygdala ROI's were then extracted. The inter-rater reliability for the extracted glucose metabolic rate is highly significant, with intra-class correlations between two independent raters being ≥ 0.97. We found that subjects with lower levels of glucose metabolism in the right amygdala report less dispositional negative affect on the PANAS scale (r's= 0.41 and 0.56 in separate samples). These findings indicate that individual differences in resting glucose metabolism in the amygdala are present and that they predict dispositional negative affect among depressed subjects.

In a small sample of 12 normal subjects, we (Irwin et al. 1996) have been able to examine the relation between the magnitude of MR signal change in the amygdala in response to aversive compared with neutral pictures and dispositional negative affect on the PANAS scale. We correlated the average value of the voxels with the maximum Student's *t* from the left and right amygdala with dispositional negative affect. There was a robust correlation, such that subjects showing the least increase in signal intensity in the right amygdala reported the lowest levels of dispositional negative affect. The findings from the fMRI and PET studies of amygdala function indicate that individual differences in both tonic activation and phasic activation in response to aversive stimuli predict the intensity of dispositional negative affect.

Emotion regulation:
A key component of affective style

One of the key components of affective style is the capacity to regulate negative emotion and specifically to decrease the duration of negative affect once it arises. We have suggested in several articles that the connections between the PFC and amygdala play an important role in this regulatory process (Davidson 1998a; Davidson and Irwin 1999; Davidson et al. 2000b). In two recent studies, we (Jackson et al. 2003; Larson et al. 1998) examined relations between individual differences in prefrontal activation asymmetry and the emotion-modulated startle. In both studies, we presented pictures from the *International Affective Picture System* (Lang et al. 1995) while acoustic startle probes were presented and the EMG-measured blink response from the orbicularis oculi muscle region was recorded (see Sutton et al. 1997a for basic methods). Startle probes were presented both during the slide exposure and at various latencies following the offset of the pictures, on separate trials. We interpreted startle magnitude during picture exposure as providing an index related to the peak of emotional response, whereas startle magnitude following the *offset* of the pictures was taken to reflect the automatic recovery from emotional challenge. Used in this way, startle probe methods can potentially provide new information on the time course of emotional responding. We expected that individual differences during actual picture presentation would be less pronounced than individual differences following picture presentation, since an acute emotional stimulus is likely to pull for a normative response across subjects whereas individuals are more likely to differ once the stimulus has terminated. Similarly, we predicted that individual differences in prefrontal asymmetry would account for more variance in predicting magnitude of recovery (i.e., startle magnitude post-stimulus) than in predicting startle magnitude during the stimulus. Our findings in each study were consistent with our predictions and indicated that subjects with greater right-sided prefrontal activation showed a larger blink magnitude following the offset of the negative stimuli, after the variance in blink magnitude *during* the negative stimulus was partialled out. Measures of prefrontal asymmetry did not reliably predict startle magnitude during picture presentation. The findings from these studies are consistent with our hypothesis and indicate that individual differences in prefrontal asymmetry are associated with the time course of affective responding, particularly the recovery following emotional challenge. In a related study, we found that subjects with greater baseline levels of left prefrontal activation are better able to voluntarily suppress negative affect (see Jackson et al. 2000a,b). Moreover, using functional MRI, we have demonstrated that, when subjects are instructed to voluntarily regulate their negative emotion, reliable bilateral changes in amygdala MR signal intensity are found (Schaefer et al. 2000) and that the magnitude of MR signal decrease in the amygdala during instructions to down-regulate negative affect are predicted by increased MR signal in the ventromedial prefrontal cortex (Urry et al. 2003).

The findings from these studies indicate that individual differences in prefrontal activation may play an important role in emotion regulation. Individuals who report less dispositional negative affect and more dispositional positive affect may be those individuals who have increased facility at regulating negative affect and, specifically, in modulating the intensity of negative affect once it has been activated.

Plasticity in the central circuitry of emotion

The circuitries that underlie emotion regulation, in particular, the amygdala and prefrontal cortex, have been targets of intensive study of plasticity (see Davidson et al. 2000a for extensive discussion). In a series of elegant studies in rats, Meaney and his colleagues (Francis and Meaney 1999) have demonstrated that an early environmental manipulation in rats – frequency of maternal licking/grooming and arched-back nursing – produces a cascade of biological changes in the offspring that shape the central circuitry of emotion and, consequently, alter the animal's behavioral and biological responsivity to stress. For example, the offspring of mothers high in licking and grooming show increased central benzodiazepine receptor densities in various subnuclei of the amygdala as well as in the locus ceruleus (LC), increased α_2 adrenoreceptor density in the LC and decreased CRH receptor density in the LC (Caldji et al. 1998). In other research, Meaney and co-workers have reported that rats exposed to high licking/grooming mothers exhibited a permanent increase in concentrations of receptors for glucocorticoids in both the hippocampus and the prefrontal cortex (Liu et al. 1997; Meaney et al. 1988, 1996). All of these changes induced by early maternal licking/grooming and related behavior involve alterations in the central circuitry of emotion that results in decreased responsiveness to stress later in life.

These findings in animals raise the possibility that similar effects may transpire in humans. There are clearly short-term changes in brain activation that are observed during voluntary emotion regulation, as noted above. Whether repeated practice in techniques of emotion regulation lead to more enduring changes in patterns of brain activation is a question that has not yet been answered in extant research. There are limited data available that indicate that cognitive behavioral therapy for certain disorders (e.g., obsessive compulsive disorder; simple phobia) produces changes in regional brain activity that are comparable to those produced by medication (Baxter et al. 1992; Paquette et al. 2003; Goldapple et al. 2004).

What is largely absent are data on plastic changes in the brain that might be produced by the practice of methods specifically designed to increase positive affect, such as meditation. In a recent study, we examined changes in brain electrical activity and immune function following an eight-week training program in mindfulness meditation (Davidson et al. 2003). In this study, subjects were randomly assigned to a meditation group or a wait-list control group and each of these groups was tested before and after the eight-week training program, as well as four months following the end of the program. We found that subjects in the meditation group showed significantly larger increases in left-sided anterior activation compared with their counterparts in the control group. Subjects received an influenza vaccine just after the eight-week program was completed and we found that influenza antibody titers were significantly higher in the meditators compared with the controls. Most remarkably, we observed that those subjects who showed the largest magnitude change in brain activity also showed the largest increase in antibody titers (see Fig. 6). In a very recent study, we (Lutz et al. 2004) tested long-term meditation practitioners and studied changes in brain electrical signals induced by the practice of specific forms of meditation. The adepts were compared with a control group who were interested in learning to meditate and were taught the practices and had one week to practice prior to the laboratory assessment. We found remarkably large increases in gamma band activity and synchrony across large-scale cortical regions in the gamma band in the adepts when they were meditating compared with the novices. These findings suggest that training procedures designed explicitly to facilitate well-being result in demonstrable and predictable changes in brain and immune function.

The Dalai Lama himself has raised this question in his recent book, *The Art of Happiness* (Dalai Lama and Cutler 1998), where he explains that "The systematic training of the mind – the cultivation of happiness, the genuine inner transformation by deliberately selecting and focusing on positive mental states and challenging negative mental states –is possible because of the very structure and function

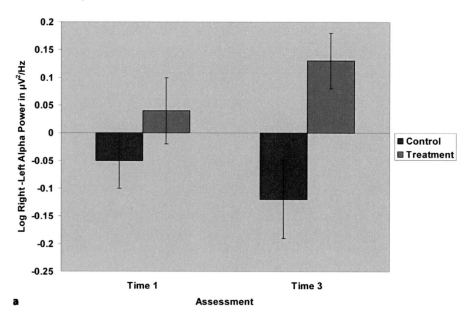

a   Assessment

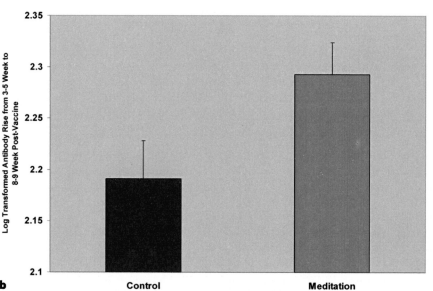

b   Control   Meditation

Fig. 6 a–b. Top panel illustrates pre- and post-training brain electrical asymmetry measures from subjects in the meditation and control groups. Note that positive numbers denote greater left-sided activation and negative numbers denote greater right-sided activation. The middle panel illustrates differences between the meditation and control groups in antibody titers in response to influenza vaccine. Errors bars in the top two plots denote standard error of the mean. The bottom panel illustrates the relation between pre-to-post test increases in left-sided activation and rise in antibody titers to influenza vaccine among subjects in the meditation group only (N=25 for meditation group; N=16 for control group; from Davidson et al. 2003).

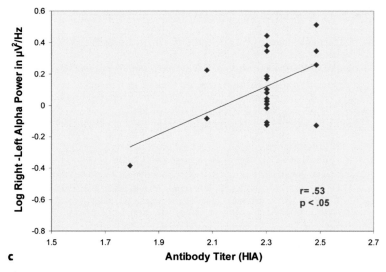

Fig. 6 c.

of the brain… But the wiring in our brains is not static, not irrevocably fixed. Our brains are also adaptable" (pp. 44–45).

Collectively the findings reviewed in this chapter indicate that there are neural reflections of positive affective styles and that these affective styles are associated with a pattern of peripheral biological function that reflects a more positive health profile. It is our conjecture that such positive affective styles are also associated with higher levels of empathy and compassion, since these are characteristics that are described as being strongly present in meditation adepts who have spent a considerable time training these neural circuits. Future research should be directed at the impact of strategies for transforming affective style on brain function and behaviors that reflect empathy and compassion.

Acknowledgment

This research was supported by NIMH grants MH43454, MH40747, P50-MH52354, P50-MH61083, NIA grant PO1-AG021079 and an NIMH training grant T32-MH1893. Portions of this chapter appeared in Davidson (2004).

References

Abercrombie HC, Schaefer SM, Larson CL, Oakes TR, Holden JE, Perlman SB, Krahn DD, Benca RM, Davidson RJ (1998) Metabolic rate in the right amygdala predicts negative affect in depressed patients. NeuroReport 9: 3301–3307

Adolphs R, Damasio H, Tranel D, Damasio AR (1995) Fear and the human amygdala. J Neurosci 15: 5879–5891

Adolphs R, Damasio H, Tranel D, Damasio AR (1996) Cortical systems for the recognition of emotion in facial expressions. J Neurosci 16: 7678–7687

Adolphs R, Tranel D, Damasio AR (1998) The human amygdala in social judgment. Nature 393, 470–474

Adolphs R, Gosselin F, Buchanan TW, Tranel D, Schyns P, Damasio AR (2005) A mechanism for impaired fear recognition after amygdala damage. Nature, 433: 68–72

Aggleton JP (1993) The contribution of the amygdala to normal and abnormal emotional states. Trends Neurosci 16: 328–333

Ahern GL, Schwartz GE (1985) Differential lateralization for positive and negative emotion in the human brain: EEG spectral analysis. Neuropsychologia 23: 745–755

Angrilli A, Mauri A, Palomba D, Flor H, Birbaumer N, Sartori G. di Paola F (1996) Startle reflex and emotion modulation impairment after a right amygdala lesion. Brain 119: 1991–2000

Baxter LR, Schwartz JM, Bergman KS, Szuba MP, Guze BH, Mazziota JC, Alazraki A, Selin CE, Ferng HK, Munford P, Phelps ME (1992) Caudate glucose metabolic rate changes with both drug and behavior therapy for obsessive-compulsive disorder. Arch Gen Psychiat 49: 681–699

Bechara A, Damasio AR, Damasio H, Anderson SW (1994) Insensitivity to future consequences following damage to human prefrontal cortex. Cognition 50: 7–15

Bechara A, Tranel D, Damasio H, Adolphs R, Rockland C, Damasio AR (1995) Double dissociation of conditioning and declarative knowledge relative to the amygdala and hippocampus in humans. Science 269: 1115–1118

Bechara A, Tranel D, Damasio H, Damasio AR (1996) Failure to respond autonomically to anticipated future outcomes following damage to prefrontal cortex. Cereb Cortex 6: 215–225

Bechara A, Damasio H, Tranel D, Damasio AR (1997) Deciding advantageously before knowing the advantageous strategy. Science 275: 1293–1295

Bechara A, Damasio H, Damasio AR, Lee GP (1999) Different contributions of the human amygdala and ventromedial prefrontal cortex to decision-making. J Neuroscie 19: 5473–5481

Birbaumer N, Grodd W, Diedrich O, Klose U, Erb E, Lotze M, Schneider F, Weiss U, Flor H (1998) fMRI reveals amygdala activation to human faces in social phobics. NeuroReport 9: 1223–1226

Breiter HC, Etcoff NL, Whalen PJ, Kennedy WA, Rauch SL, Buckner RL, Strauss MM, Hyman SE, Rosen BR (1996a) Response and habituation of the human amygdala during visual processing of facial expression. Neuron 17: 875–887

Breiter HC, Rauch SL, Kwong KK, Baker JR, Weisskoff RM, Kennedy DN, Kendrick AD, Davis TL, Jiang A, Cohen, MS, Stern CE, Belliveau JW, Baer L, O'Sullivan RL, Savage CR, Jenike MA, Rosen BR (1996b) Functional magnetic resonance imaging of symptom provocation in obsessive-compulsive disorder. Arch Gen Psychiat 53: 595–606

Broks, P, Young AW, Maratos EJ, Coffey PJ, Calder AJ, Isaac CL, Mayes AR, Hodges JR, Montaldi D, Cezayirli E, Roberts N, Hadley D (1998) Face processing impairments after encephalitis: amygdala damage and recognition of fear. Neuropsychologia 361: 59–70

Buchel C, Morris J, Dolan RJ, Friston KJ (1998) Brain systems mediating aversive conditioning: An event-related fMRI study. Neuron 20: 947–957

Cacioppo JT, Gardner WL (1999) Emotion. Ann Rev Psychol 50: 191–214

Cahill L, McGaugh JL (1998) Mechanisms of emotional arousal and lasting declarative memory. Trends Neurosci 21: 273–313

Cahill L, Weinberger NM, Roozendaal B, McGaugh JL (1999) Is the amygdala a locus of "conditioned fear"? Some questions and caveats. Neuron 23: 227–228

Calder AJ, Young A.W, Rowland D, Perrett DI, Hodges JR, Etcoff NL (1996) Facial emotion recognition after bilateral amygdala damage: differentially severe impairment of fear. Cogn Neuropsychol 135: 699–745

Caldji C, Tannenbaum B, Sharma S, Francis D, Plotsky PM, Meaney MJ (1998) Maternal care during infancy regulates the development of neural systems mediating the expression of fearfulness in the rat. Proc Natl Acad Sci USA 95: 5335–5340

Carver CS, White TL (1994) Behavioral inhibition, behavioral activation and affective responses to impending reward and punishment: the BIS/BAS scales. J Personal Soc Psychol 67: 319–333

Dalai Lama, Cutler HC (1998) The art of happiness. New York: Riverhead Books.

Davidson RJ (1995) Cerebral asymmetry, emotion and affective style. In: Davidson RJ, Hugdahl K (eds) Brain asymmetry. Cambridge, MA: MIT Press, pp. 361–387

Davidson RJ (1998a) Affective style and affective disorders: perspectives from affective neuroscience. Cogn Emotion 12: 307–320

Davidson RJ (2004) Well-being and affective style: Neural substrates and biobehavioral correlates. Phil Trans Roy Soc (London) 359: 1395–1411

Davidson RJ, Fox NA (1989) Frontal brain asymmetry predicts infants' response to maternal separation. J Abnormal Psychol 98: 127–131

Davidson RJ, Tomarken AJ (1989) Laterality and emotion: an electrophysiological approach. In: Boller F, Grafman J (eds) Handbook of neuropsychology. Vol. 3. Amsterdam: Elsevier, pp. 419–441

Davidson RJ, Sutton SK (1995) Affective neuroscience: the emergence of a discipline. Curr Opin Neurobiol 5: 217–224

Davidson RJ, Irwin W (1999) The functional neuroanatomy of emotion and affective style. Trends Cogn Sci 3: 11–21

Davidson RJ, Chapman JP, Chapman LP, Henriques JB (1990) Asymmetrical brain electrical activity discriminates between psychometrically-matched verbal and spatial cognitive tasks. Psychophysiology 27: 238–543

Davidson RJ, Kalin NH, Shelton SE (1992) Lateralized effects of diazepam on frontal brain electrical asymmetries in rhesus monkeys. Biol Psychiat 32: 438–451

Davidson RJ, Kalin NH Shelton SE (1993) Lateralized response to diazepam predicts temperamental style in rhesus monkeys. Behav Neurosci 107: 1106–1110

Davidson RJ, Coe CC, Dolski, Donzella B (1999) Individual differences in prefrontal activation asymmetry predicts natural killer cell activity at rest and in response to challenge. Brain Behav Immunity 13: 93–108

Davidson RJ, Jackson DC, Kalin NH (2000a) Emotion, plasticity, context, and regulation: perspectives from affective neuroscience. Psychol Bull 126: 890–909

Davidson RJ, Putnam KM, Larson CL. (2000b) Dysfunction in the neural circuitry of emotion regulation – a possible prelude to violence. Science 289: 591–594

Davidson RJ, Jackson DC, Larson CL (2000c) Human electroencephalography. In: Cacioppo JT, Bernston GG, Tassinary LG (eds) Principles of psychophysiology New York: Cambridge University Press. pp.27–52

Davidson RJ, Marshall JR, Tomarken AJ, Henriques JB (2000d) While a phobic waits: regional brain electrical and autonomic activity in social phobics during anticipation of public speaking. Biol Psychiat 47: 85–95

Davidson RJ, Pizzagalli D, Nitschke JB, Putnam KM (2002) Depression: perspectives from affective neuroscience. Ann Rev Psychol 53: 545–574

Davidson R.J, Kabat-Zinn J, Schumacher J, Rosenkrantz, M, Muller D, Santorelli SF, Urbanowski F, Harrington A, Bonus K, Sheridan JF (2003) Alterations in brain and immune function produced by mindfulness meditation. Psychosomat Med 65: 564–570

Davis M (1992) The role of the amygdala in fear-potentiated startle: Implications for animal models of anxiety. Trends Pharmacol Sci 13: 35–41

Davis M, Whalen PJ (2001) The amygdala: vigilance and emotion. Mol Psychiat 6: 13–34

Fanselow MS, LeDoux JE (1999) Why we think plasticity underlying Pavlovian fear conditioning occurs in the basolateral amygdala. 23: 229–232

Francis D, Meaney MJ (1999) Maternal care and development of stress responses. Curr Opin Neurobiol 9: 128–134

Frijda NH (1994) Emotions are functional, most of the time. In: Ekman P, Davidson RJ (eds) The nature of emotion: fundamental questions New York: Oxford University Press, pp. 112–122

Gainotti G (1972) Emotional behavior and hemispheric side of lesion. Cortex 8: 41–55

Giese-Davis J, Spiegel D (2003) Emotional expression and cancer progression. In: Davidson RJ, Scherer K, Goldsmith HH (eds) Handbook of affective neuroscience. New York: Oxford University Press, pp. 1053–1082

Gold PW, Goodwin FK, Chrousos GP (1988) Clinical and biochemical manifestations of depression: relation to the neurobiology of stress. New Engl J Med 314: 348–353

Goldapple K, Segal Z, Garson C, Lau M, Bieling P, Kennedy S, Mayberg H (2004) Modulation of cortical-limbic pathways in major depression: Treatment-specific effects of cognitive therapy. Arch Gen Psychiat 61: 34–41

Goldman-Rakic PS (1996) The prefrontal landscape: implications of functional architecture for understanding human mentation and the central executive. Phil Transactions of the Roy Soc London 351: 1445–1453

Goldman-Rakic PS (2000) Localization of function all over again. NeuroImage 11: 451–457

Hamann SB, Ely TD, Hoffman JM, Kilts CD (2002) Ecstasy and agony: activation of the human amygdala in positive and negative emotion. Psychol Sci 13:135–41

Hugdahl K (1998) Cortical control of human classical conditioning: Autonomic and positron emision tomography data. Psychophysiology 35: 170–178

Hugdahl K, Beradi A, Thompson WL, Kosslyn SM, Macy R, Baker DP, Alpert NM, LeDoux JE (1995) Brain mechanisms in human classical conditioning: a PET blood flow study. NeuroReport 6: 1723–1728

Irwin W, Davidson RJ, Lowe MJ, Mock BJ, Sorenson JA, Turski PA (1996) Human amygdala activation detected with echo-planar functional magnetic resonance imaging. NeuroReport 7: 1765–1769

Jackson DC, Malmstadt J, Larson CL , Davidson RJ (2000a) Suppression and enhancement of emotional responses to unpleasant pictures. Psychophysiology 37: 515–522

Jackson DC, Burghy CA, Hanna A, Larson, CL, Davidson RJ(2000b) Resting frontal and anterior temporal EEG assymmetry predicts ability to regulate negative emotion. Psychophysiology, 37: S50

Jackson DC, Mueller CJ, Dolski IV, Dalton KM, Nitschke JB, Urry HL, Rosenkranz MA, Ryff CD, Singer BH, Davidson RJ (2003) Now you feel it, now you don't: Frontal brain electrical asymmetry and individual differences in emotion regulation. Psychol Sci 14: 612–617.

Jones NA, Fox NA (1992) Electroencephalogram asymmetry during emotionally evocative films and its relation to positive and negative affectivity. Brain Cogni 20: 280–299

Kalin NH, Larson CL, Shelton SE, Davidson RJ (1998) Asymmetric frontal brain activity, cortisol, and behavior associated with fearful temperament in Rhesus monkeys. Behav Neurosci 112: 286–292

Kalin, NH, Shelton SE,Davidson RJ (2000) Cerebrospinal fluid corticotropin-releasing hormone levels are elevated in monkeys withpatterns of brain activity associated with fearful temperament. Biol Psychiat 47: 579–585

Kalin N.H, Shelton SE, Davidson R J, Kelley AE (2001) The primate amygdala mediates acute fear but not the behavioral and physiological components of anxious temperament. J Neurosci 21: 2067–2074

Kang DH, Davidson RJ, Coe CL, Wheeler RW, Tomarken AJ, Ershler WB (1991) Frontal brain asymmetry and immune function. Behav Neurosci 105: 860–869

Kiecolt-Glaser JK, Glaser R (1981) Stress and immune function in humans. In: Ader R, Felten DL, Cohen N (eds) Psychoneuroimmunology San Diego, CA: Academic Press, pp. 849–867

Kosslyn SM, Koenig O (1992) Wet mind: the new cognitive neuroscience. New York: Free Press

LaBar KS, Gatenby JC, LeDoux JE, Phelps EA (1998) Human amygdala activation during conditioned fear acquisition and extinction – a mixed-trial fMRI study. Neuron 205: 937–945

Lang PJ (1995) The emotion probe: studies of motivation and attention. Am Psychol 50: 372–385

Lang PJ, Bradley MM, Cuthbert BN (1995) International affective picture system IAPS: technical manual and affective ratings. Gainseville, FL: The Center for Research in Psychophysiology University of Florida

Larson C., Sutton SK, Davidson R.J (1998) Affective style, frontal EEG asymmetry and the time course of the emotion-modulated startle. Psychophysiology 35: S52

LeDoux JE (1996) The emotional brain: the mysterious underpinnings of emotional lift. New York: Simon & Schuster

Levenson RW (1994) Human emotion: a functional view. In: Ekman P, Davidson RNJ (eds) The nature of emotion: fundamental questions New York: Oxford University Press, pp. 123–126

Liu D, Diorio J, Tannenbaum B, Caldji C, Francis D, Freedman A (1997) Maternal care, hippocampal glucocoricoid receptors, and hypothalamic-pituitary-adrenal responses to stress. Science 277: 1659–1662

Lutz A, Greischar, LL, Rawlings NB, Ricard M, Davidson RJ (2004) Long-term meditators self-induce high-amplitude synchrony during mental practice. Proc Natl Acad Sci USA 101: 16369–16373

Meaney MJ, Aitken DH, van Berkel C, Bhatnagar S, Sapolsky RM (1988) Effect of neonatal handling on age-related impairments associated with the hippocampus. Science 239: 766–768

Meaney MJ, Bhatnagar S, Larocque S, McCormick CM, Shanks N, Sharman S, Smythe J, Viau V, Plotsky, PM (1996) Early environment and the development of individual differences in the hypothalamic-pituitary-adrenal stress response. In: Pfeffer CR (eds) Severe stress and mental disturbance in children Washington, D.C.: American Psychiatric Press pp. 85–127

Miller A,Tomarken AJ (2001) Task-dependent changes in frontal brain asymmetry: effects of incentive cues, outcomes expectancies, and motor responses. Psychophysiology 38: 500–511

Morris JS, Frith CD, Perrett DI, Rowland D, Young AW, Calder AJ, Dolan RJ (1996) A differential neural response in the human amygdala to fearful and happy facial expressions. Nature 383: 812–815

Paquette V, Levesque J, Mensour B, Leroux JM, Beaudoin G, Bourgouin P, Beauregard M (2003) Change the mind and you change the brain: effects of cognitive-behavioral therapy on the neural correlates of spider phobia. NeuroImage 18: 401–409

Phillips ML, Young AW, Senior C, Brammer M, Andrews C, Calder AJ, Bullmore ET, Perrett DI, Rowland D, Williams SCR, Gray JA, David AS (1997) A specific neural substrate for perceiving facial expressions of disgust. Nature 389: 495–498

Rauch SL, van der Kolk BA, Fisler RE, Alpert NM, Orr SP, Savage CR, Fischman AJ, Jenike MA, Pitman RK (1996) A symptom provocation study of posttraumatic stress disorder using positron emission tomography and script-driven imagery. Arch Genl Psychiat 535: 380–387

Rauch SL, Savage CR, Alpert NM, Fischman AJJenike MA (1997) A study of three disorders using positron emission tomography and symptom provocation. Biol Psychiat 42: 446–452

Robinson RG, Starr LB, Price TR (1984) A two year longitudinal study of mood disorders following stroke: prevalence and duration at six months follow-up. Brit J Psychiat 144: 256–262

Rolls ET (1999) The brain and emotion. New York: Oxford University Press

Rosenkranz MA, Jackson DC, Dalton KM, Dolski I, Ryff CD, Singer BH, Muller D, Kalin NH, Davidson RJ (2003) Affective style and in vivo immune response: Neurobehavioral mechanisms. Proc Natl Acad Sci USA 100: 11148–11152

Sackeim HA, Greenberg MS, Weiman AL, Gur RC, Hungerbuhler JP, Geschwind N (1982) Hemispheric asymmetry in the expression of positive and negative emotions: Neurologic evidence. Arch Neurol 39: 210–218

Schaefer SM, Jackson DC, Davidson RJ, Aguirre GK, Kimberg DY, Thompson-Schill SL (2002) Modulation of amygdalar activity by the conscious regulation of negative emotion. J Cogn Neurosci 14: 913–921

Schneider F, Gur RE, Alavi A, Seligman MEP, Mozley LH, Smith RJ, Mozley PD, Gur RC (1996) Cerebral blood flow changes in limbic regions induced by unsolvable anagram tasks. Am J Psychiat 153: 206–212

Scott SK, Young AW, Calder AJ, Hellawell DJ, Aggleton JP, Johnson M (1997) Impaired auditory recognition of fear and anger following bilateral amygdala lesions. Nature 385: 254–257

Shackman AJ, Maxwell JS, Skolnick AJ, Schaefer HS, Davidson RJ (2003) Exploiting individual differences in the prefrontal asymmetry of approach-related affect: hemodynamic, electro-encephalographic, and psychophysiological evidence. Program No. 444.6. 2003 Abstract Viewer/Itinerary Planner. Washington, DC: Society for Neuroscience, online

Sobotka S.S, Davidson RJ, Senulis JA (1992) Anterior brain electrical asymmetries in response to reward and punishment. Electroencephalogr Clin Neurophysio 83: 236–247

Sutton SK, Davidson RJ (1997a) Prefrontal brain asymmetry: a biological substrate of the behavioral approach and inhibition systems. Psychol Sci 8: 204–210

Sutton SK, Davidson RJ, Donzella B, Irwin W, Dottl DA (1997b) Manipulating affective state using extended picture presentation. Psychophysiology 34: 217–226

Sutton SK, Ward RT, Larson CL, Holden JE, Perlman SB, Davidson RJ (1997) Asymmetry in prefrontal glucose metabolism during appetitive and aversive emotional states: An FDG-PET study. Psychophysiology 34: S89

Taylor SE (1991) Asymmetrical effects of positive and negative events: The mobilization-minimization hypothesis. Psychol Bull 110: 67–85

Tomarken AJ, Davidson RJ, Wheeler RE, Doss RC (1992) Individual differences in anterior brain asymmetry and fundamental dimensions of emotion. J Personal Soc Psychol 62: 676–687

Urry HL, van Reekum CM, Johnstone T, Thurow ME, Burghy CA, Mueller CJ, Davidson RJ (2003) Neural correlates of voluntarily regulating negative affect. Program No. 725.18. 2003 Abstract Viewer/Itinerary Planner. Washington, DC: Society for NeuroScience online

Urry HL, Nitschke JB, Dolski I, Jackson DC, Dalton KM, Mueller CJ, Rosenkranz MA, Ryff CD, Singer BH, Davidson RJ (2004) Making a life worth living: neural correlates of well-being. Psycholl Sci 15: 367–372

Vrana SR, Spence EL, Lang PJ (1988) The startle probe response: A new measure of emotion? J Abnormal Psychol 97: 487–49

Watanabe M (1996) Reward expectancy in primate prefrontal neurons. Nature 382: 629–632

Watson D, Clark LA, Tellegen A (1988) Developmental and validation of brief measures of positive and negative affect: the PANAS scales. J Personal Soc Psychol 54: 1063–1070

Whalen PJ, Rauch SL, Etcoff NL, McInerney SC, Lee MB, Jenike MA (1998) Masked presentations of emotional facial expressions modulate amygdale activity without explicit knowledge. J Neurosci 19: 411–418

Wilson DS (1994) Adaptive genetic variation and human evolutionary psychology. Ethnol Sociobiol 154: 219–235

Zald DH, Pardo JV (1997) Emotion, olfaction and the human amygdala: Amygdala activation during aversive olfactory stimulation. Proc Natl Acad Sci USA 94: 4119–4124

Zald DH, Lee JT, Fluegel KW, Pardo JV (1998) Aversive gustatory stimulation activates limbic circuits in humans. Brain 121: 1143–1154

Zalla T, Koechlin E, Pietrini P, Basso G, Aquino P, Sirigu A, Grafman J (2000) Differential amygdala responses to winning and losing: a functional magnetic resonance imaging study in humans. Eur J Neurosci 12: 1764–1770

Cognitive Psychology of Moral Intuitions

Daniel Kahneman[1] *and Cass R. Sunstein*[2]

There have been profound changes in the psychological analysis of moral sentiments over the last few decades, from a conception of morality as a system of
abstract rules that can be understood and internalized (Kohlberg 1969) to a view
that emphasizes moral emotions and moral intuitions that are not anchored in
reasons e.g., Greene and Haidt 2002; Haidt 2001; Rozin et al. 1999). In this brief
essay, we sketch an analysis of moral intuitions that builds on the new work and
relates it to a general approach to the study of intuitive thought. We suppose that
the same cognitive machinery generates attitudes, judgments, beliefs and actions
in moral domains and in other domains, and that the moral domain is distinctive because it involves a special attitude: the complex of emotion, beliefs and
response tendencies that define *indignation*.

Rozin et al. (1999) suggested that there are three variants of indignation: anger,
disgust and contempt. The first two, but perhaps not the third, are strongly associated with altruistic punishment. We are mainly concerned here with the variant
of indignation that involves anger. For an example, imagine that you see a bully
beat up a weakling without any provocation. You will respond with indignation.
Like other attitudes, indignation has three related constituents:
1) an *emotion*, which, confusingly, is also named indignation
2) a set of *beliefs and judgments* about the reprehensible action
3) a *response tendency* to administer punishment to the guilty actor even at some
   cost to oneself (economists call this a readiness to provide altruistic punishment, because it occurs even in the absence of any expectation of further interaction with the target of punishment).

Like other intentional states, indignation can be explained in two quite different
ways: by referring to reasons or to psychological causes. As you see the bully as-
saulting his victim, you are likely to be aware of a reason for your emotion: the
action violates an accepted (and in your view, justified) social rule that prohibits
unprovoked aggression. The categorization of the action provides a reason for indignation, a reason that the observer expects other objective observers to endorse.
Classical analyses of moral development were much concerned with people's abil-

[1] Department of Psychology, Princeton University Green Hall - Washington Road, Princeton NJ
08544, USA; e-mail: kahneman@princeton.edu
[2] University of Chicago, Dept. of Political Science and the College, 1111 East 60th Street, Chicago,
IL 60637, USA; e-mail: csunstei@midway.uchicago.edu

Changeux et al.
Neurobiology of Human Values
© Springer-Verlag Berlin Heidelberg 2005

ity to marshal reasons for their judgments; the reasons were often understood as causing those judgments.

The view that has gained currency in recent years is quite different. In this view, indignation is like a fear of spiders. One does not fear spiders because they are dangerous: one just fears them. Because people tend to attribute their reactions to the objects that evoke these reactions, the feared spider is perceived as a dangerous spider. However, the perception of dangerousness is not the reason for the fear or even its cause; both the fear and the perception are symptoms of an uncontrolled reaction to spiders. Many people who are afraid of spiders know that their fear is objectively groundless and lacks a reason. Haidt (2001) described the equivalent state in the moral domain as "moral dumbfounding," the experience of strong moral reactions for which no adequate reason comes to mind. Indignation, we suggest, is often not caused by reasons, and people can be dumbfounded when they are asked to explain why they are indignant. In fact, some puzzling outcomes, in both politics and law, are a product of indignation that is intense and hard to justify (Sunstein 2005). Moral dumbfounding finds its mirror image in moral numbness, in which people are not indignant even though they have reason to be, and know they do.

The two-system model of the mind

Consider the expression, "$17 \times 24 = ?$" For the great majority of people, the correct answer to the question will come to mind only if it is produced by a voluntary mental activity, which involves deliberate application of a rule, requires several steps of computation, storage and retrieval and takes a significant amount of time. For contrast, consider the word "vomit." For the great majority of people, disgust will come to mind in a completely involuntary process, which is produced very quickly by a process that is itself unconscious: one is aware only of its outcome. The two examples represent different families of cognitive processes.

The ancient idea that cognitive processes can be partitioned into two main families – traditionally called "intuition" and "reason" – is now widely embraced under the general label of dual-process theories (Chaiken and Trope 1999; Sloman 1996). Dual-process theories come in many forms, but all distinguish cognitive operations that are quick and associative from others that are slower, more reflective, and frequently more calculative (Gilbert 1999). We adopt the generic labels System 1 and System 2 from Stanovich and West (1999). These terms may suggest the image of autonomous homunculi, and there is in fact evidence that the two systems correspond to different locations in the brain, but we do not suggest that the two systems are independent. We use "systems" as a label for collections of processes that are distinguished by their speed, their controllability, and the contents on which they operate (see Table 1).

Although System 1 is more primitive than System 2, it is not necessarily less capable. On the contrary, complex cognitive operations eventually migrate from System 2 to System 1 as proficiency and skill are acquired. A striking demonstration of the intelligence of System 1 is the ability of chess masters to perceive

Table 1. Two cognitive systems

System 1 (Intuitive)	System 2 (Reflective)
Automatic	Controlled
Effortless	Effortful
Associative	Deductive
Rapid	Slow
Opaque process	Self-aware
Skilled	Rule-following

the strength or weakness of chess positions instantly. For those experts, pattern matching has replaced effortful serial processing. Some people are widely taken to be moral experts as well, and it should be clear that pattern matching occurs in the moral domain. Indignation is often a result.

In the particular dual-process model that we assume, System 1 quickly proposes intuitive answers to judgment problems as they arise, and System 2 monitors the quality of these proposals, which it may endorse, correct, or override. The judgments that are eventually expressed are called intuitive if they retain the hypothesized initial proposal without much modification. The roles of the two systems in determining stated judgments depend on features of the task and of the individual, including the time available for deliberation (Finucane et al. 2000) and the respondent's mood (Isen et al.1988; Bless et al.1996) and intelligence (Stanovich and West 1999). Without time for deliberation, for example, indignation can be extremely intense; when people have time to reflect, their reaction diminishes. And when System 1 is not indignant, and people are morally numb, deliberation can heighten moral concern and sometimes produce indignation (though this can take a great deal of heavy lifting on the part of System 2). We assume that System 1 and System 2 can be concurrently active, that automatic and controlled cognitive operations compete for the control of overt responses, and that deliberate judgments are likely to remain anchored on initial impressions. Our views in these regards are similar to other dual-process models (Epstein 1994; Gilbert 1989, 2002; Sloman 1996).

The properties listed in Table 1 are shared by the system that produces intuitive thoughts and by the perceptual system. Intuitive judgments appear to occupy a position – perhaps corresponding to evolutionary history – between the automatic operations of perception and the deliberate operations of reasoning. Unlike perception, however, the operations of System 1 are not restricted to the processing of current stimulation. Like System 2, the operations of System 1 deal with stored concepts as well as with precepts and can be evoked by language. This view of intuition suggests that the vast store of scientific knowledge available about perceptual phenomena can be a source of useful hypotheses about the workings of intuition. The strategy of drawing on analogies from perception is applied in the following section.

A defining property of intuitive thoughts is that they come to mind spontaneously, like percepts. The technical term for the ease with which mental contents come to mind is *accessibility* (Higgins 1996). To understand intuition, we must understand why some thoughts are accessible and others are not.

Some attributes are more accessible than others, both in perception and in judgment. Attributes that are routinely and automatically produced by the perceptual system or by System 1, without intention or effort, have been called *natural assessments* (Tversky and Kahneman 1983). For example, experimental evidence shows that, when a perceiver is exposed to a set of objects of the same general kind (e.g., a set of lines of different sizes), attributes of a prototypical member of the set (e.g., the average length of the lines) are computed effortlessly and automatically. Other attributes (e.g., the total length of the lines) are not accessible; they can only be assessed by a deliberate and quite laborious computation. Thus, average length is a natural assessment but total length is not. Kahneman and Frederick (2002) compiled a partial list of these natural assessments. In addition to physical properties such as size, distance, and loudness, the list includes more abstract properties such as similarity, causal propensity, surprisingness, affective valence, and mood.

The evaluation of stimuli as good or bad is a particularly important natural assessment. The evidence, both behavioral (Bargh 1997; Zajonc 1998) and neurophysiological (e.g., LeDoux 2000), is consistent with the idea that the assessment of whether objects are good (and should be approached) or bad (should be avoided) is carried out quickly and efficiently by specialized neural circuitry. A remarkable experiment reported by Bargh (1997) illustrates the speed of the evaluation process and its direct link to approach and avoidance. Participants were shown a series of stimuli on a screen and instructed to respond to each stimulus as soon as it appeared by moving a lever that blanked the screen. The stimuli were affectively charged words, some positive (e.g., LOVE) and some aversive (e.g., VOMIT), but this feature was irrelevant to the participant's task. Half the participants responded by pulling the lever toward themselves, half responded by pushing the lever away. Although the response was initiated within a fraction of a second, well before the meaning of the stimulus was consciously registered, the emotional valence of the word had a substantial effect. Participants were relatively faster in pulling a lever toward themselves (approach) for positive words and relatively faster pushing the lever away when the word was aversive. The tendencies to approach or avoid were evoked by an automatic process that was not under conscious voluntary control. Several psychologists have commented on the influence of this primordial evaluative system (here included in System 1) on the attitudes and preferences that people adopt consciously and deliberately (Zajonc 1998; Kahneman et al. 1999; Slovic et al 2002; Epstein 2003). The implicit attitude test, designed to measure racial and other biases, finds that people show an automatic bias against African-Americans, older people, and others, even when they are unaware of it and wish to be unbiased (Nosek et al. 2002).

To complete this sketch of the operations of System 1, we introduce a process of *attribute substitution* that shapes many judgments and choices. The concept was introduced by Kahneman and Frederick (2002) as a basic mechanism to explain the basic results of heuristic judgment. They proposed that the reduction of complex tasks to simpler operations, which characterizes such judgments, is achieved by an operation in which an individual assesses "a specified *target attribute* of a judgment object by substituting another property of that object – the *heuristic attribute* – which comes more readily to mind" (p. 53). In the moral domain, as we show later, individuals charged with the task of determining the severity of a punishment appear to solve this difficult problem by consulting the intensity of their outrage.

Several of the processes we have discussed are involved in the explanation of a study reported by Miller and McFarland (1986), in which respondents determined the appropriate compensation for a man who was shot in the arm during the robbery of a grocery store. Some respondents were told that the robbery happened at the victim's regular store. Other respondents were told that the victim was shot in a store that he visited for the first time, because his usual store happened to be closed that day. The two versions obviously differ in poignancy, because the counterfactual "undoing" of an unusual event comes more easily to mind than the undoing of a normal occurrence. The difference of poignancy translated into a difference of $100,000 in the median award judged appropriate for the two cases. The participants in this experiment apparently answered the difficult question of appropriate question by mapping onto a scale of dollars their answer to a simple question: how much were they emotionally touched by the story.

It is unlikely that the respondents deliberately chose to provide this large compensation for poignancy. Indeed, when respondents were presented with both versions of the robbery story and asked whether a compensation board should make different awards in the two cases, 90% thought it should not. In the terms of the present discussion, the emotion-anchored process that produced the initial awards is dominated by System 1. The requirement to compare two questions evokes a much more complex activity, here attributed to System 2, which identifies the distinctive element that separates the two versions and is unable to find any moral justification for different awards. This can be seen as an instance of "moral dumbfounding" (Haidt 2001), in which a strong intuition exists that cannot be anchored in rules that the person consciously accepts.

Outrage and harm

We have studied the operation of moral judgments in the domain of punitive damage awards (Kahneman et al. 1998; Sunstein et al. 1998). One of our hypotheses, couched in the language of the present treatment, was that the setting of such awards is mediated by an outrage heuristic.

Participants drawn from a jury roll in Texas were shown vignettes of legal cases in which a plaintiff had suffered a personal injury while using a product. For example, one of the scenarios concerned a child who had been burned when his pajamas caught fire as he was playing with matches. The pajamas were made of fabric that was not adequately fire-resistant, and the defendant firm had been aware of the problem. For some of the scenarios, alternative versions were constructed that differed in the severity of harm. In the high-harm version of the pajamas case, for example, the child was "severely burned over a significant portion of his body and required several weeks in hospital and months of physical therapy." In the low-harm version, "his hands and arms were badly burned and required professional medical treatment for several weeks". Participants were told that the plaintiff had already been awarded compensatory damages. One group of respondents indicated whether *punitive damages* were appropriate, and if so, in what amount. Another group rated the *outrageousness* of the defendant's behavior. In a subsequent re-analysis of this study, Kahneman and Frederick (2002) also obtained ratings of the *severity of the harm* suffered in each of the 14 vignettes. Lawsuits were not mentioned in these descriptions of harm. The same basic design was replicated twice, varying the size of the defendant firm.

The results supported the conclusion that assessments of punitive damages (the target attribute in this study) were mediated by an *outrage heuristic*. In the analysis offered by Kahneman and Frederick (2002), the outrage associated with each case was estimated by the product of the product of the average ratings of outrageousness and of harm. The correlations (over 14 vignettes) between the estimate of outrage and mean punitive damages were 90 in one of the firm-size conditions and .94 in the other.

The role of actual harm as a determinant of outrage in this experiment is of interest as a potential case of moral dumbfounding. The legally recognized distinction between murder and attempted murder is a salient example of the issue we raise. Consider the following scenarios:

1) A wishes B dead but does nothing about it
2) A tries to kill B and fails by chance
3) A tries to kill B and succeeds

It is difficult to justify a moral distinction between the last two cases. Indeed, it is safe to assume that people are asked to judge the outrageousness of the *actions*, there will be no difference. But punitive intent reflects the emotional intensity of the response to the event, and the emotion evidently depends on the harm that actually occurred. In the terms of the present analysis, the severity of punishment reflects the intensity of an emotional reaction in System 1. Punishments that are determined in this manner are expected to be crudely retributive, which is what we observe. Note that we are not claiming that it is impossible to defend the distinction, drawn by the criminal law, between murder and attempted murder. There may be good reasons for drawing that distinction. What we are arguing is that the distinction is not caused by those reasons; it is caused by the fact that moral intuitions, automatic and uncontrolled, are different in the two cases.

Indignation and the definition of harm

Indignation is evoked in an observer "by an agent who, intentionally and without provocation or adequate reason, causes a victim to suffer harm." This highly abstract statement could be read as a rule that specifies appropriate reasons for indignation. Instead, we take the statement as an attempt to describe the rules that govern the elicitation of indignation in the observer's System 1. As in many other situations, the use of a language of reason to describe processes that reason does not govern creates ambiguity and the risk of confusion. Indeed, every word in the statement changes its meaning across the two contexts, and the meaning that applies to the description must be a concept that System 1 is able to process.

An example is the notion of harm. Here again, we can offer a defining statement: "Harm is a loss relative an entitlement. An individual's entitlements are governed by rules and expectations that are shared by the community." This statement obviously covers many different kinds of harm, from physical injury to loss of property, and also to loss of reputation, depending on the nature of the entitlement that is violated. Here again, the statement sounds like a rule. It also appears to lead us nowhere, because now we must ask, what does System 1 know about entitlements? System 2 might, of course, have a theory of one or another kind, but that theory may or may not map onto moral intuitions. For System 1, the answer is that an entitlement is a socially endorsed *normal state*, also called a *reference state*, relative to which losses are defined. A reference state is an expectation that a valued stated will be maintained. A valued state is defined by the fact that deviations from it produce positive or negative affect. Now we are making some progress, because System 1 (and the perceptual system) are quite capable of setting up expectations and in detecting deviations from them, and System 1 also produces emotional responses to changes.

The intuitive notion of entitlement was explored in a study of the moral attitudes to the behavior of firms in the market (Kahneman et al. 1986). Residents of two Canadian cities evaluated the fairness of various unilateral actions that a firm might take to set the terms of its transactions with customers, employees or tenants. Table 2 presents examples that convey the general trends of the results, by attaching the labels Fair and Unfair to several sets of actions. Each set may be viewed as an experiment, intended to isolate a critical factor in judgments of unfairness. Actions that are grouped together were always evaluated by different respondents.

A surprisingly simple model, growing out of a conception of what counts as harm, adequately summarizes those intuitive judgments. In this model, the relations between a firm and its transactors (customers or employees) are regulated by a *reference transaction*, which specifies a reference profit for the firm and reference terms (price, wage, rent) for the transactors. The firm has the power to impose a change of these terms by posting new prices or by setting new wages; it is also free to maintain the reference terms of trade as circumstances change. Actions that abuse this power will be called unfair.

The reference transaction specifies what the transactors "have a right to expect" in their dealings with the firm. According to widespread moral intuitions, both sides have an informal entitlement to the terms of that transaction. The reference terms often correspond to the status quo. Indeed, the linking of entitlements to the reigning status quo is a powerful mechanism that transforms *is* into *ought*. However, states other than the status quo can be designated as the relevant reference. For example, a firm that labels a price as discounted or a part of the worker's income as a bonus reserves the right to limit its transactors' entitlement to the "regular" prices and wages.

A few critical distinctions are made. The first is between actions that the firm takes to protect its profits from an exogenous threat and actions that it initiates

Tab e 2. Evaluation of the fairness of various unilateral actions

A1	Unfair	A hardware store raises the price of snow shovels during a spring blizzard.
A2	Fair	A store maintains the price of a good when its costs drop, thus increasing its profits.
B1	Unfair	An employer who is doing well cuts wages by 5%, because many unemployed workers would be willing to work at the lower rate.
B2	Fair	An employer who is doing poorly cuts wages by 5%, to avoid or diminish losses.
C1	Fair	The worker replacing an employee who quit is paid less for doing the same job.
C2	Fair	An employer changes his line of business (from painting to landscaping), retains the same workers but reduces their wages.
D1	Unfair	A landlord raises the rent of a sitting tenant after learning that the tenant has found a good job in the area and will not want to move.
D2	Fair	Facing increasing costs, a landlord raises the rent of a sitting tenant who lives on a fixed income and may be forced to move.
E1	Unfair	A grocery store with a large stock of peanut butter raises its price immediately when it learns that wholesale prices have risen.
E2	Fair	A landlord who rents apartments in two identical buildings charges higher rent for one of them, because a more costly foundation was required for that building.
F1	Unfair	An employer cuts wages by 7% in a period of high unemployment and no inflation.
F2	Fair	An employer raises wages by 5% in a period of high unemployment and 12% inflation.
	Unfair	A car dealer raises the price of a popular model above list price when a shortage develops.
	Fair	A car dealer eliminates a discount on a popular car model.

to increase its profit at someone's expense. The principal constraint imposed by the intuitive principles of fairness is that a firm cannot fairly achieve a gain by exploiting its power to impose a loss on its transactors. This is the sense in which transactors have an entitlement to the reference terms. However, the entitlement is not unconditional. The firm is also entitled to its reference profit – the profit it achieves in the reference trasaction – and this entitlement trumps the transactors' entitlement. When its reference profit is threatened, it is not unfair for a firm to protect itself by passing on losses to transactors; the rules of fairness do not require the firm to absorb any part of an exogenously imposed loss. A second distinction is between losses and failures to gain. The rules of fairness do not require the firm to share with its transactors any gains that it obtains, whether by good fortune or by its efforts. For example, a large majority of respondents in a subsequent study believed that a firm has no obligation to pass on any reduction of wholesale prices to its customers. Of course, a firm that shares its gains with its customers or employees will be admired. But a firm that fails to do so will not evoke indignation.

As we might expect of judgments that arise from intuitions rather than from a rule, the moral judgments in Table 2 are specified at a strikingly low level of abstraction,. For example, the reference price for a can of peanut butter appears to be linked to the price at which it was bought, rather than to the price at which it will be replaced (E2), and the firm's obligation not to cut wages evidently protects only the nominal wage rather than the workers' real income (F1 vs F2). The entitlement to the reference terms is also personal: it protects the current tenant or employee, but not their replacements.

It seems safe to predict that the factors that determine the fairness of market transactions will also prove to be critical in the broader context of social justice. In law as well as morality, the issue of affirmative action provides an instructive example. Hardly anyone will think it fair to achieve the goals of affirmative action by a policy in which current employees would lose their jobs to make room for replacements that would improve the representativeness of the work force. In fact, the Supreme Court of the United States has vindicated this intuition, suggesting that affirmative action programs are unacceptable when they impose losses on existing workers (Wygant v. Jackson Board of Education, 476 US 267, 1986). On the other hand, a policy that discriminates among new hires is much more acceptable, because it appears to create only gainers and non-gainers, but no losers. As this example illustrates, the distinction between actual losses and opportunities foregone, however fragile after reflection, is psychologically and morally real. Indeed, the distinction appears to be categorical: the coefficient of loss aversion is much higher in the moral domain than in the domain of individual choice. System 2, of course, raises serious questions about the distinction between actual losses and opportunities foregone, but it is often silenced in the face of moral indignation.

Moral Framing

A framing effect is said to occur when two extensionally equivalent statements evoke different judgments or preferences when presented singly, yet appear transparently equivalent when shown together (Tversky and Kahneman 1981). Framing effects arise because statements that are extensionally equivalent may nevertheless evoke different associations and different emotional responses. A cold cut described as 90% fat-free is more attractive than if it is described as 10% fat, and it is more likely to be purchased. Framing effects are a manifestation of the associative and emotional processes of System 1. There have been several demonstrations of framing effects in the domain of moral judgment.

Thomas Schelling has offered a compelling example, one that illustrates the difficulties that may arise when outcomes are evaluated as changes relative to an arbitrary reference level. Schelling reports asking his students to evaluate a tax policy that would allow a larger child exemption to the rich than to the poor (Schelling 1984). Many listeners will immediately respond, as Schelling's students did, that this proposal is outrageous. The intuition that leads to this quick conclusion is a simple and attractive rule: no social allocation of gains or losses should be more favorable to the rich than to the poor. Schelling next points out to his students that the default case in the standard tax table is a childless family, with special adjustments for families with children. Of course, the existing tax schedule could be rewritten with a family with two children as the default case. Childless families would then pay a surcharge. Should this surcharge be as large for the poor as for the rich? Of course not.

Schelling's example is a framing effect, in which an inconsequential aspect of the statement of the problem appears to control moral judgments. The intuitions that are evoked by formulating the problem in terms of exemptions and surcharges are incoherent. Neither intuition survives when the inconsistency is pointed out. It soon becomes apparent that the only formulation of the tax problem that avoids arbitrariness is a complete table that determines the after-tax income of families for every level of pre-tax income and number of children. This description in terms of final states is superior precisely because it avoids powerful intuitions that can be manipulated by the framing of the problem. Unfortunately, there are no clear moral intuitions left to resolve the problem when it is properly framed. In this case as well as in others the only potent moral intuitions apply to changes and to differences, not to states.

From money we turn to legal questions involving the valuation of injuries to health and to an experiment in which the same difference between two states of health was caused to be coded either as a loss or as a gain. The experiment was concerned with lay assessments of appropriate monetary compensation for the pain and suffering associated with personal injuries, such as "losing mobility in one knee for four years" (McCaffery et al. 1995). Separate samples of respondents were given different "jury instructions" describing the thought experiment they should conduct to determine fair compensation. One of the instructions suggested a positive choice between two desirable options. The respondents were instructed to imagine that the victim had very recently suffered the injury and was now offered a choice between a complete and immediate cure and an amount of money. Fair compensation was to be set at the highest amount for which the victim would still prefer the cure. In contrast, the selling instruction required the respondent to assume that the victim considered an ex ante proposition to accept the injury in conjunction with a payment of money. Fair compensation was to be set at the lowest payment for which the victim would have accepted the offer.

The difference between health and injury is coded as a gain in the former case and as a loss in the latter: this is the pattern of an endowment effect. In terms of final states, of course, the two versions of the problem are not distinguishable. As expected, the average judgment of fair compensation was about twice as high with the selling than with the choice instruction. This is also a framing effect, but, unlike Schelling's example, the people who are susceptible to this effect rarely acknowledge its existence. When the participants in each experimental condition were shown the instruction given to the other group, they thought both instructions were fair and did not notice that they were likely to evoke discrepant responses. The legal system typically uses a version of the buying instruction, but without a great deal of reflective thinking about why that instruction should be preferred, and creative lawyers are able to frame the problem so as to ensure that a selling instruction comes before the jury in a way that produces predictably higher dollar awards.

Moral framing has been demonstrated in the important context of obligations to future generations (see Frederick 2003), a much-disputed question of morality, politics, and law (Revesz 1999). To say the least, the appropriate discount rate for those yet to be born is not a question that most people have pondered, and hence their judgments are highly susceptible to different frames. From a series of surveys, Maureen Cropper and her coauthors (1994) suggest that people are indifferent between saving one life today and saving 45 lives in 100 years. They make this suggestion on the basis of questions asking people whether they would choose a program that saves "100 lives now" or a program that saves a substantially larger number "100 years from now." It is possible, however, that people's responses depend on uncertainty about whether people in the future will otherwise die (perhaps technological improvements will save them?); other ways of framing the same problem yield radically different results (Frederick 2003). For example, most people consider "equally bad" a single death from pollution next year and a single death from pollution in 100 years. This finding implies no preference for members of the current generation. The simplest conclusion is that people's moral intuitions about obligations to future generations are very much a product of framing effects (for a similar result, see Baron 2000).

The same point holds for the question whether government should consider not only the number of "lives" but also the number of "life years" saved by regulatory interventions. If the government focuses on life years, a program that saves children will be worth far more attention that a similar program that saves senior citizens. Is this immoral? People's intuitions depend on how the question is framed (see Sunstein 2004). People will predictably reject an approach that would count every old person as "worth less" than what every young person is worth. But if people are asked whether they would favor a policy that saves 105 old people or 100 young people, many will favor the latter, in a way that suggests a willingness to pay considerable attention to the number of life years at stake.

Framing effects present a large difficulty for the achievement of coherent judgments and preferences. The normal process of comprehension takes a given message to a state of the world, but the correspondence of messages and states is not one-to-one. Ambiguity arises when a single message is compatible with multiple states of the world. Framing effects arise when a single state of the world may be described in multiple ways, and when a relevant response is description-dependent. Thus, the avoidance of framing effects requires a search through the set of descriptions that are extensionally equivalent to the original message. Unfortunately, the human mind is not equipped to solve this problem. People are therefore given no warning, when confronted with a particular version of Schelling's tax puzzle, that there is an alternative version of the same problem to which they would have responded differently. In the context of punitive damage awards and valuation of environmental amenities, we have found that incoherence in moral judgments is predictable (Sunstein et al. 2002). We leave to others the possible implications of these observations for the notion of reflective equilibrium.

Agency, omission and brutal commission

To say the least, there has been much discussion of whether and why the distinction between acts and omissions might matter for morality, law, and policy. In one case, for example, a patient might ask a doctor not to provide life-sustaining equipment, thus ensuring the patient's death. In another case, a patient might ask a doctor to inject a substance that will immediately end the patient's life. Many people seem to have a strong moral intuition that the failure to provide life-sustaining equipment, and even the withdrawal of such equipment, is acceptable and legitimate but that the injection is morally abhorrent. And indeed, American constitutional law reflects judgments to exactly this effect. People have a constitutional right to withdraw equipment that is necessary to keep them alive, but they have no constitutional right to physician-assisted suicide (see Washington v. Glucksberg 1997, pp. 724–25). But what is the morally relevant difference?

It is worth considering the possibility that the act-omission distinction is rooted in System I and is in some cases very hard to defend in principle. The moral puzzles arise when life, or a clever interlocutor, comes up with a case in which there is no morally relevant distinction between acts and omissions but when moral intuitions strongly suggest that there must be such a difference. As an example, consider the vaccination question. Many people show an omission bias, favoring inaction over statistically preferable action (Baron and Ritov 1994). The persistent acceptance of withdrawal of life-saving equipment, alongside persistent doubts about euthanasia, may be another demonstration of the point.

Compare the dispute over two well-known problems in moral philosophy (see Thomson 1986, pp. 94–116). These problems do not involve the act-omission distinction; no omission is involved. But the problems implicate closely related concerns. The first, called the trolley problem, asks people to suppose that a runaway trolley is headed for five people who will be killed if the trolley continues on its current course. The question is whether you would throw a switch that would move the trolley onto another set of tracks, killing one person rather than five. Most people would throw the switch. The second, called the footbridge problem, is the same as that just given, but with one difference: the only way to save the five is to throw a stranger, now on a footbridge that spans the tracks, into the path of the trolley, killing that stranger but preventing the trolley from reaching the others. Most people will not kill the stranger; in fact they are indignant at the suggestion that they ought to do so. But what is the difference between the two cases, if any? A great deal of philosophical work has been done on this question, much of it trying to suggest that our firm intuitions can indeed be defended in principle.

Without engaging these arguments, let us suggest the possibility of a simpler answer. As a matter of principle, there may or may not be a difference between the two cases. But people's different reactions are based on automatic moral intuitions that condemn the throwing of the stranger but support the throwing of the switch. As a matter of intuition, it is worse to throw a human being in the path of a trolley than to throw a switch that (indirectly?) leads to a death. People also struggle heroically, and by reference to System 2, to rescue their intuitions and to establish that the two cases are genuinely different in principle, whether or not this is so, But System 1, and indignation about brutal acts of commission, are responsible for the underlying intuitions.

Consider a suggestive experiment designed to see how the human brain responds to the two problems (Greene et al. 2001). The authors do not attempt to answer the moral questions in principle, but they find "that there are systematic variations in the engagement of emotions in moral judgment," and that brain areas associated with emotion are far more active in contemplating the footbridge problem than in contemplating the trolley problem. An implication of the authors' finding is that human brains are hard-wired to distinguish between bringing about a death "up close and personal" and doing so at a distance. It follows that acts, especially brutal acts, would be far more likely to produce reactions from the brain areas associated with emotions than omissions that cause identical harms. Compare the case of fear, where an identifiable region of the brain makes helpfully immediate but not entirely reliable judgments (LeDoux 1998), in a way that suggests a possible physical location for some of the operations of System I. So too, we think, in the context of morality, politics, and law (Greene and Haidt 2002). A clear implication involves moral numbness. Many omissions do not trigger indignation on the part of System 1 but might well be subject to moral criticism from the standpoint of System 2, if only it can become sufficiently active.

Conclusion

Moral intuitions operate in much the same way as other intuitions do; what makes the moral domain as distinctive is its foundations in the emotions, beliefs, and response tendencies that define indignation. System 1 is typically responsible for indignation; System 2 may or may not provide an override. Moral dumbfounding and moral numbness are often products of moral intuitions that people are unable to justify. We have argued that an understanding of indignation helps to explain the operation of the outrage heuristic, the centrality of harm, the role of reference states, moral framing, and the act-omission distinction. We have also suggested that, because of how indignation operates, it is extremely difficult for people to achieve coherence in their moral intuitions.

The intuitions described here play an important role in multiple domains, including families, workplaces, sporting events, and religious organizations. But as many of our examples suggest, they also influence the decisions of legal and political institutions. Such institutions usually aspire to be deliberative, and to pay close attention to System 2, but even in deliberative institutions, System 1 can make some compelling demands.

References

Bargh JA (1997) The automaticity of everyday life. In: Wyer Jr, RS (ed) The automaticity of everyday life: advances in social cognition. Vol. 10. Mahwah, NJ: Erlbaum, pp.1–61

Baron J (2000) Can we use human judgments to determine the discount rate? Risk Analysis 20: 861–868

Baron J, Ritov I (1994) Reference points and omission bias. Org Behav Human Decision Proc 59: 475–498

Bless H, Clore GL, Schwarz N, Golisano V, Rabe C, Wolk M (1996) Mood and the use of scripts: Does a happy mood really lead to mindlessness? J Personal Soci Psychol 71: 665–679

Chaiken S, Trope Y (eds) (1999) Dual-process theories in social psychology. New York: Guilford Press

Cropper ML, Aydede SK, Portney PR (1994) Preferences for life-saving programs: How the public discounts time and age. J Risk Uncertainty 8: 243–265

Epstein S (1994) Integration of the cognitive and psychodynamic unconscious. Am Psychol 49: 709–724.

Epstein S (2003) Cognitive-experiential self-theory of personality. In: Millon T, Lerner MJ (eds) Comprehensive handbook of psychology. Vol. 5. Personality and social psychology. Hoboken, NJ: Wiley, pp. 159–184

Finucane ML, Alhakami A, Slovic P, Johnson SM (2000) The affect heuristic in judgments of risks and benefits. J Behav Decision Making 13: 1–17

Frederick S (2003) Measuring intergenerational time preference: Are future lives valued less? J Risk Uncertainty 26,: 39–53

Gilbert DT (1989) Thinking lightly about others: Automatic components of the social inference process. In: Uleman J, Bargh JA (eds) Unintended thought. Englewood Cliffs, NJ: Prentice-Hall, pp. 189–211

Gilbert DT (1999) What the mind's not. In: Chaiken S, Trope Y (eds) Dual process theories in social psychology. New York: Guilford

Gilbert DT (2002) Inferential correction. In: Gilovich T, Griffin D, Kahneman D (eds) Heuristics and biases. New York: Cambridge University Press, pp. 167–184

Greene J, Haidt J (2002) How (and where) does moral judgment work? Trends Cogn Sci 6: 517–523

Greene JD, Cohen JD (2004) For the law, neuroscience changes nothing and everything. Phil Trans Rol Socif London B, (Special Issue on Law and the Brain), 359: 1775–1785

Greene J, Somerville RB, Nystrom LE, Darley JM, Cohen JD (2001) An fMRI investigation of emotional engagement in moral judgment. Science 293: 2105–2108

Haidt J (2001) The emotional dog and its rational tail: A social intuitionist approach to moral judgment. Psychol Rev 108: 814–834

Higgins ET (1996) Knowledge activation: accessibility, applicability, and salience. In: Higgins ET, Kruglanski A (eds) Social psychology: Handbook of basic principles> New York: Guilford Press, pp. 133–168

Isen, AM, Nygren TE, Ashby FG (1988).Influence of positive affect on the subjective utility of gains and losses: It is just not worth the risk. J Personal Soc Psychol 55: 710–717

Kahneman D, Frederick S (2002) Representativeness revisited: attribute substitution in intuitive judgment. In: Gilovich T, Griffin D, Kahneman D (eds) Heuristics and biases. New York: Cambridge University Press, pp 49–81

Kahneman D, Knetsch J, Thaler R (1986) Fairness as a constraint on profit seeking: Entitlements in the market. Am Econ Rev 76: 728–741

Kahneman D, Schkade DA, Sunstein CR (1998) Shared outrage and erratic awards: The psychology of punitive damages. J Risk Uncertainty 16: 49–86

Kahneman D, Ritov I, Schkade D (1999) Economic preferences or attitude expressions? An analysis of dollar responses to public issues. J Risk Uncertainty 19: 220–242

Kohlberg Lawrence (1969) "Stage and sequence: The cognitive-developmental approach to socialization." In: Goslin DA (ed) Handbook of socialization theory and research. Chicago: Rand McNally, pp. 347–480

LeDoux J (1996) The emotional brain: the mysterious underpinning of emotional life. Touchstone

LeDoux JE (2000) Emotion circuits in the brain. Ann Rev Neurosci 23: 155–184

McCaffery EJ, Kahneman D, Spitzer ML (1995) Framing the jury: cognitive perspectives on pain and suffering awards. Virginia Law Rev 81: 1341–1420

Miller DT, McFarland C (1986) Counterfactual thinking and victim compensation: a test of norm theory. Personal Soc Psychol Bull 12: 513–519

Nosek B, Banaji M, Greenwald A (2002) Harvesting implicit group attitudes and beliefs from a demonstration website. Group Dynamics: Theory Res Practice 6: 101–116

Revesz R (1999) Environmental regulation, cost-benefit analysis, and the discounting of human lives. Columbia Law Rev 99: 941–1017

Rozin P, Lowery L, Imada S (1999) The CAD triad hypothesis: a mapping between three moral emotions (contempt, anger, disgust) and three moral codes (community, autonomy, and divinity). J Personali Soc Psychol 76: 574–586

Schelling TC (1984) Choice and consequence: Perspectives of an errant economist. Cambridge MA: Harvard University Press

Sloman SA (1996) The empirical case for two systems of reasoning. Psychol Bull 119: 3–22

Slovic P, Finucane M, Peters E, MacGregor DG (2002) The affect heuristic. In: Gilovich T, Griffin D, Kahneman D (eds) Heuristics and biases. New York: Cambridge University Press, pp. 347–420

Stanovich KE, West RF (1999) Discrepancies between normative and descriptive models of decision making and the understanding/acceptance principle. Cogn Psychol 38: 349–385

Sunstein CR (2004) Lives, life-years, and willingness to pay. Columbia Law Rev 104: 205–252

Sunstein C (2005) Moral heuristics. Behav Brain Sci, in press

Sunstein C, Kahneman D, Schkade D (1998) Assessing punitive damages. Yale Law J 107: 2071–2153

Sunstein CR, Kahneman D, Schkade D, Ritov I (2002) Predictably incoherent judgments. Stanford Law Rev 54: 1153–1216

Thomson JJ (1986) Rights, restitution, and risk: essays in moral theory. Cambridge, MA: Harvard University Press

Tversky A, Kahneman D (1981) The framing of decisions and the psychology of choice. Science 211: 453–458

Tversky A, Kahneman D (1983) Extensional vs. intuitive reasoning: The conjunction fallacy in probability judgment. Psychol Rev 90: 293–315

Washington V. Glucksberg (1997) West's Supreme Court Reporter 521:702–789

Wygant V. Jackson Board of Education, 476 US 267 (1986)

Zajonc R.B (1998) Emotions. In: Gilbert DT, Fiske ST, Lindzey G (eds) Handbook of social psychology. 4th Ed., Vol. 1. New York: Oxford University Press, pp 591–632

Mirror neuron: a neurological approach to empathy

Giacomo Rizzolatti[1,3] *and Laila Craighero*[2]

Summary

Humans are an exquisitely social species. Our survival and success depend critically on our ability to thrive in complex social situations. But how do we understand others? Which are the mechanisms underlying this capacity?

In the present essay we discuss a general neural mechanism ("mirror mechanism") that enables individuals to understand the meaning of actions done by others, their intentions, and their emotions, through activation of internal representations coding motorically the observed actions and emotions.

In the first part of the essay we will show that the mirror mechanism for "cold" actions, those devoid of emotional content, is localized in the parieto-frontal cortical circuits. These circuits become active both when we do an action and when we observe another individual doing the same action. Their activity allows the observer to understand the "what" of an action.

We will show, then, that a "chained" organization of motor acts plus the mirror mechanism enable the observer to understand the intention behind an action (the "why" of an action) by observing the first motor act of an action.

Finally, we will discuss some recent data showing that the mirror mechanism localized in other centers, like the insula, enables the observer to understand the emotions of others. We will conclude briefly discussing whether these biological data allow inferences about moral behavior.

Introduction

"How selfish soever man may be supposed, there are evidently some principles in his nature, which interest him in the fortune of others, and render their happiness necessary to him, though he derives nothing from it except the pleasure of seeing it." This famous sentence by Adam Smith (1759), which so nicely describes our empathic relation with others, contains two distinct concepts.

---

[1] Dipartimento di Neuroscienze, Sezione di Fisiologia, via Volturno, 3, Università di Parma, 43100, Parma, Italy;

[2] Dip. SBTA, Sezione di Fisiologia Umana, via Fossato di Mortara, 17/19, Università di Ferrara, 44100 Ferrara, Italy;

[3] Corresponding author: Giacomo Rizzolatti, e-mail: giacomo.rizzolatti@unipr.it

Changeux et al.
Neurobiology of Human Values
© Springer-Verlag Berlin Heidelberg 2005

The first is that individuals are endowed with a mechanism that makes them share the "fortunes" of others. By observing others, we enter in an "empathic" relation with them. This empathic relation concerns not only the emotions that others feel but also their actions. "The mob, when they are gazing at a dancer on the slack rope, naturally writhe and twist and balance their own bodies, as they see him do, and as they feel that they themselves must do if in his situation."(Smith 1759).

The second idea is that, because of our empathy with others, we are compelled to desire their happiness. If others are unhappy, we are also unhappy, because the other's unhappiness somehow intrudes into us.

According to Smith, the way in which we enter into empathic relation may be voluntary ("As we have no immediate experience of what other men feel, we can form no idea of the manner in which they are affected, but by conceiving what we ourselves should feel in the like situation"), but also, as shown in the above-cited example of the "dancer," may be automatically triggered by the observation of the behavior of others.

The aim of this essay is to discuss the existence of a neural mechanism, resembling that described by Adam Smith, that puts the individual in empathic contact with others. This mechanism – the mirror mechanism – enables the observer to understand the actions of others, the intention behind their actions, and their feelings. In a short "coda," we will discuss the validity of the second idea of Adam Smith, the obligatory link between our happiness and that of others.

Action understanding

Humans are social beings. They spend a large part of their time observing others and trying to understand what they are doing and why. Not only humans but also apes and monkeys have a strongly developed interest in others.

How are actions recognized? The traditional view is that action recognition is based exclusively on the visual system. The understanding of an action done by another individual depends on the activity of the higher order visual areas and, in particular, of those of the superior temporal sulcus, where there are neurons selectively activated by biological motions (Perrett et al. 1989; Carey et al. 1997; Allison et al. 2000; Puce and Perrett 2003).

Another hypothesis is that an action is recognized when the observed action activates, in the observer's brain, an analogous motor representation. The observer does not execute that action, because control mechanisms prevent its overt occurrence, but the evoked motor representation ("motor knowledge") allows him to understand the meaning of what he saw (Rizzolatti et al. 2001).

It is important to note that the two hypotheses are not in contraposition. Rather, they describe two different ways in which an action may be understood. The "visual" hypothesis describes a "third person" relation between the observer and the observed action. The action, albeit recognized in its general meaning, is not understood in all its implications, because it does not enter into the semantic motor network of the observing individual as well as in his/her private knowledge of what doing that action means. "Visual" understanding is similar to that a robot, able to differentiate an action from another, may have, or humans have when they see a bird flying or a dog barking (see below). In contrast, the "motor" hypothesis describes the "first person" understanding of what the individual is seeing. The observed action enters into the observer's motor representation and recalls his/her similar experiences when doing that action. It is an empathic recognition that makes the observer share the experience of the action agent.

Mirror Neuron System and Action Understanding

The strongest evidence in favor of the "motor hypothesis" is represented by mirror neurons. These neurons, originally found in the monkey ventral premotor cortex (area F5), are active both when the monkey *does* a particular action and when it *observes* another individual doing a similar action. The mirror neurons do not respond to object presentation. Similarly, they do not respond to the sight of an agent mimicking actions or performing non-object-directed gestures. Mirror neurons have also been described in the monkey parietal lobe (Fogassi et al. 1998; Gallese et al. 2002).

The mirror mechanism appears to be a mechanism particularly well suited for imitation. Imitation, however, appeared only late in evolution. Monkeys, which have a well-developed mirror system, lack this capacity and even apes have it only in a rather rudimentary form (see Tomasello and Call 1997; Visalberghi and Fragaszy 2002).

The properties of monkey mirror neurons also indicate that this system initially evolved not for imitation. Mirror neurons typically show a good congruence between the visual actions they respond to and the motor responses they code, yet only in a minority of them do the effective observed and effective executed actions correspond in terms of *both* goal and means for reaching the goal. Most of them code the goal of the action (e.g., grasping) but not the way in which the observed action is done. These neurons are, therefore, of little use for imitation in the proper sense, that is the capacity to imitate an action as it has been performed (Rizzolatti and Craighero 2004).

Summing up, it is very likely that the faculty for imitation developed on the top of the mirror system. However, its initial basic function was not imitation but enabling an individual to understand actions done by others (see Rizzolatti et al. 2001).

Evidence in favor of the notion that mirror neurons are involved in action understanding comes from a recent series of studies in which mirror neurons were tested in experimental conditions in which the monkey could understand the meaning of an occurring action but had no visual information about it. The rationale of these studies was the following: if mirror neurons mediate action understanding, they should become active when the meaning of the observed action is understood, even in the absence of visual information.

The results showed that this is the case. In one study, F5 mirror neurons were recorded while the monkey was observing a "noisy" action (e.g., ripping a piece of paper) and then was presented with the same noise without seeing the action . The results showed that a large number of mirror neurons, responsive to the observation of noisy actions, also responded to the presentation of the sound proper of that action, alone. Responses to white noise or to the sound of other actions were absent or much weaker than responses to the preferred action. Neurons responding selectively to specific action sounds have been named "audio-visual" mirror neurons (Kohler et al. 2002).

In another study, mirror neurons were tested by introducing a screen between the monkey and the location of an object (Umiltà et al. 2001). The idea underlying the experiment was that, if mirror neurons are involved in action understanding, they should also discharge in conditions in which the monkey does not see the occurring action but has sufficient clues to create a mental representation of what the experimenter does. The monkeys were tested in four conditions: 1) the experimenter is grasping an object; 2) the experimenter is miming grasping, and 3) and 4), the monkey observes the actions of 1) and 2) but the final critical part of them (hand-object interaction) is hidden by a screen. It is known that mirror neurons typically do not fire during the observation of mimed actions. At the beginning of each "hidden " trial, the monkey was shown whether there was or was not an object behind the screen. Thus, in the hidden condition, the monkey "knew" that the object was present behind the screen and could mentally represent the action.

The results showed that more than half of the tested neurons discharged in the hidden condition, thus indicating that the monkey was able to understand the goal of the action even in the absence of the visual aspect describing the action.

In conclusion, both the experiments show that the activity of mirror neurons correlates with action understanding. The visual features of the observed actions are necessary to trigger mirror neurons only insomuch as they allow the understanding of the observed actions. If action comprehension is possible on other bases, mirror neurons signal the action even in the absence of visual stimuli.

The Mirror Neuron System in Humans

Evidence, based on single neuron recordings, of the existence of mirror neurons in humans is lacking. Their existence is, however, indicated by EEG and MEG studies, TMS experiments, and brain imaging studies (see Rizzolatti and Craighero 2004). For the sake of space, we will review here only a tiny fraction of these studies.

MEG and EEG studies showed that the desynchronization of the motor cortex found during active movements was also present during the observation of action done by others (Hari et al. 1998; Cochin et al. 1999). Recently, desynchronization of cortical rhythms was found in functionally delimited language and hand motor areas in a patient with implanted subdural electrodes, both during observation and execution of finger movements (Tremblay et al. 2004).

TMS studies showed that the observation of actions done by others determines an increase of corticospinal excitability with respect to the control conditions. This increase concerned specifically those muscles that the individuals use for producing the observed movements (e.g., Fadiga et al. 1995; Strafella and Paus 2000; Gangitano et al. 2001, 2004).

Brain imaging studies allowed the localization of the cortical areas forming the human mirror neuron system. They showed that the observation of actions done by others activates, besides visual areas, two cortical regions whose function is classically considered to be fundamentally or predominantly a motor one: the inferior parietal lobule (area PF/PFG), and the lower part of the precentral gyrus (ventral premotor cortex) plus the posterior part of the inferior frontal gyrus (IFG) (Rizzolatti et al. 1996; Grafton et al. 1996; Grèzes et al. 1998, 2003; Iacoboni et al. 1999, 2001; Nishitani and Hari 2000, 2002; Buccino et al. 2001; Koski et al. 2002, 2003; Manthey et al. 2003; Johnson-Frey et al. 2003). These two regions form the core of the mirror neuron system in humans.

Recently, an fMRI experiment was carried out to see which type of observed actions is recognized using the mirror neuron system (Buccino et al. 2004). Video-clips showing silent mouth actions done by humans, monkeys and dogs were presented to normal volunteers. Two types of actions were shown: biting and oral communicative actions (speech reading, lip-smacking, barking). Static images of the same actions were presented as a control.

The results showed that the observation of biting, regardless of whether done by a man, a monkey or a dog, determined the same two activation foci in the inferior parietal lobule, and in the *pars opercularis* of IFG and the adjacent precentral gyrus. Speech reading activated the left *pars opercularis* of IFG, whereas the observation of lip smacking activated a small focus in the right and left *pars opercularis* of IFG. Most interestingly, the observation of barking *did not* produce any mirror neuron system activation.

These results strongly support the notion that actions done by other individuals can be recognized through different mechanisms. Actions belonging to the motor repertoire of the observer are mapped on his/her motor system. Actions that do not belong to this repertoire do not excite the motor system of the observer and appear to be recognized essentially on a visual basis. Thus, in the first case, the motor activation translates the visual experience into an empathic, first person knowledge, whereas this knowledge is lacking in the second case.

Intention understanding

The data discussed above indicate that the premotor and parietal cortices of primates contain a mechanism that allows individuals to understand the actions of others. Typically, an individual observing an action done by another person not only understands what that person is doing, but also why he/she is doing it. Let us imagine a boy grasping a mug. There are many reasons why the boy could grasp it, but the observer usually is able to infer why he did it. For example, if the boy grasped the cup by the handle, it is likely that he wants to drink the coffee, while if he grasped it by the top it is more likely that he wants to place it in a new location.

The issue of whether the intention comprehension (the "why" of an observed action) could be mediated by the mirror neurons has been recently addressed in a study in which the motor and visual properties of mirror neurons of the inferior parietal lobule (IPL) were investigated (Fogassi et al.2005).

IPL neurons that discharge during active grasping were selected. Subsequently, their motor activity was studied in two main conditions. In the first, the monkey grasped a piece of food located in front of it and brought it to its mouth (eating condition). In the second, the monkey grasped an object and placed it into a container (placing condition).

The results showed that the large majority of IPL grasping neurons (about 65% of them) were significantly influenced by the action in which the grasping was embedded. Examples are shown in Figure 1. Neurons coding grasping for eating were much more common than neurons coding grasping for placing, with a ratio of two to one.

Studies in humans showed that the kinematics of the first motor act of an action is influenced by the subsequent motor acts of that action (see Jeannerod 1988). The recordings of reaching-to-grasp kinematics of the monkeys in the two experimental conditions described above confirmed these findings. Reaching-to-grasp movement followed by arm flexion (bringing the food to the mouth) was faster than the same movement followed by arm abduction (placing the food into the container). To control for whether the differential discharge of grasping neurons in eating and placing conditions were due to this difference in kinematics rather than to the action goal, a new placing condition was introduced (see Fig. 1). In this condition the monkey had to grasp a piece of food and place it into a container located near its mouth. Thus, the new place condition was identical in terms of goal to the original one but required, after grasping, arm flexion rather than arm abduction. The kinematics analysis of the reaching-to-grasp movement showed that the wrist peak velocity was fastest in the new placing condition, intermediate in the eating condition, and slowest in the original placing condition.

Neuron activity showed that, regardless of arm kinematics, the neuron selectivity remained unmodified. Neurons selective for placing in the far container showed the same selectivity for placing in the container near the monkey's mouth. Thus, it is the goal of the action that determines the motor selectivity of IPL neurons in coding a given motor act, rather than factors related to movement kinematics.

As in the premotor cortex, there are neurons in IPL that are endowed with mirror properties, discharging both during the observation and execution of the same motor act (Fogassi et al. 1998, Gallese et al. 2002). To see whether these neurons also discharge differentially during the observation of the same motor act

Fig. 1. A. Lateral view of the monkey brain showing the sector of IPL (gray shading) from which the neurons were recorded; cs, central sulcus; ips, inferior parietal sulcus. B Schematic drawing illustrating the apparatus and the paradigm used for the motor task. Left: starting position of the task. A screen prevents the monkey from seeing the target. Right: after the screen is removed, the monkey could release the hand from the starting position, reach and grasp the object to bring it to its mouth (Condition I) or to place it into a container located near the target (II) or near its mouth (III). C Activity of three IPL neurons during grasping in Conditions I and II. Rasters and the histograms are synchronized with the moment when the monkey touched the object to be grasped. Abscissa: time, bin = 20 ms; ordinate: discharge frequency. (From Fogassi et al. 2005)

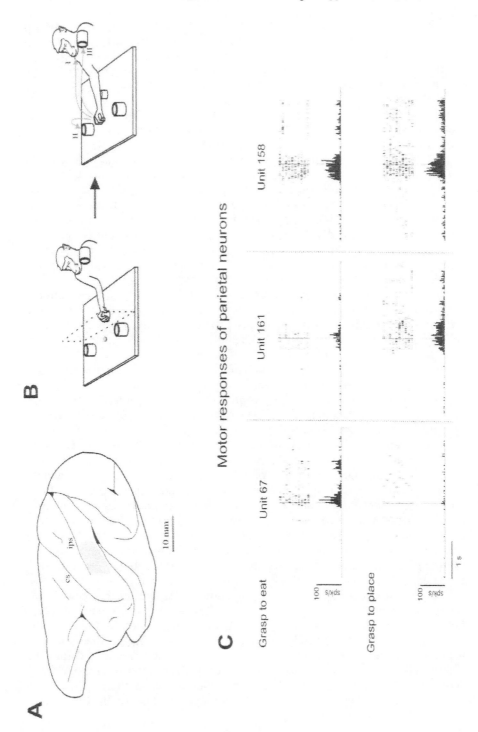

but embedded in different actions, their visual properties were tested in the same two conditions as those used for studying their motor properties. The actions were performed by an experimenter in front of the monkey. In one condition, the monkey observed the experimenter grasping a piece of food and bringing it to his mouth; in the other, the monkey observed the experimenter placing an object into the container.

The results showed that more than two-thirds of IPL neurons were differentially activated during the observation of grasping in placing and eating conditions. Examples are shown in Figure 2. Neurons responding to the observation of grasping for eating were more represented than neurons responding to the observation of grasping for placing, again with a two to one ratio.

A comparison between neuron selectivity during the execution of a motor act and motor act observation showed that the great majority of neurons (84%) have the same specificity during grasping execution and grasping observation. Thus, a mirror neuron whose discharge was higher during the observation of grasping for eating than during the observation of grasping for placing also had a higher discharge when the monkey grasped for eating than when it grasped for placing. The same was true for neurons selective for grasping for placing.

The motor and visual organization of IPL, just described, is of great interest for two reasons. First, it indicates that motor actions are organized in the parietal cortex in specific chains of motor acts; second, it strongly suggests that this chained organization might constitute the neural basis for understanding the intentions of others.

In favor of the existence of action chains in IPL, it is not only the activation of neurons coding the same motor act in one condition and not in another, but also the organization of IPL neuron-receptive fields. This organization shows that there is a predictive facilitatory link between subsequent motor acts. To give an example, there are IPL neurons that respond to the passive flexion of the forearm, have tactile receptive fields on the mouth, and in addition discharge during mouth grasping (Ferrari et al., manuscript in preparation). These neurons facilitat the mouth opening when an object touches the mouth, but also when the monkey grasps it, producing a contact between the object and the hand tactile-receptive field. Recently, several examples of predictive chained organization in IPL have been described by Yokochi et al. (2003). If one considers that a fundamental aspect of action execution is its fluidity, the link between the motor acts forming an action and the specificity of neurons coding them appears to be an optimal solution for executing an action without having pauses between the individual motor acts forming it.

The presence of chained motor organization of IPL neurons has deep implications for intention understanding. The interpretation of the functional role of mirror neurons was, as described above, that of action understanding. A motor act done by another individual is recognized when this act triggers the same set of neurons that are active during that act execution. The action-related IPL mirror neurons allow one to extend this concept. These neurons discriminate one motor act from another, thus activating a motor act chain that codes the final goal of the action. In this way the observing individual may re-enact internally the observed action and thus *predict* the goal of the observed action. In this way, the observer can "read" the intention of the acting individual.

This intention-reading interpretation predicts that, in addition to mirror neurons that fire during the execution and observation of the same motor act ("classical mirror neurons"), there should be neurons that are visually triggered by a given motor act but discharge during the execution *not* of the same motor act, but of another one that is functionally related to the former and is part of the same action chain. Neurons of this type have been previously described both in F5 (Di Pellegrino et al. 1992) and in IPL (Gallese et al. 2002) and referred to as "logically related" mirror neurons. These "logic" mirror neurons were never theoretically discussed because their functional role was not clear. The findings just discussed allow us to not only account for their occurrence but also to indicate their necessity, if the chain organization is at the basis of intention understanding.

While the mechanism of intention understanding just described appears to be rather simple, it would be more complex to specify how the selection of a particular chain occurs. After all, what the observer sees is just a hand grasping a piece of food or an object.

There are various factors that may determine this selection. The first is the context in which the action is executed. In the study described above, the clue for possible understanding of the intention of the acting experimenter was either the presence of the container (placing condition) or its absence (eating condition). The second factor that may intervene in chain selection is the type of object that the experimenter grasped. Typically, food is grasped in order to be eaten. Thus, the observation of a motor act directed towards food is more likely to trigger grasping-for-eating neurons than neurons that code grasping for other purposes. This food-eating association is, of course, not mandatory but could be modified by other factors.

One of these factors is the standard repetition of an action. Another is, as mentioned before, the context in which the action is performed. Context and object type were found to interact in some neurons. For example, some neurons that selectively discharged during the observation of grasping for eating also discharged, although weakly, during the observation of grasping for placing when the object to be placed was food, but not when it was a solid. It was as if the eating chain was activated, although slightly, by food in spite of the presence of a contextual clue indicating that placing was the most likely action. A few neurons, instead of showing an intermediate discharge when the nature of the stimulus (food) and context conflicted, decreased their discharge with time when the same action was repeated. It was as if the activity of the placing chain progressively inhibited the activity of neurons of the eating chain.

Understanding "other minds" constitutes a special domain of cognition. Developmental studies clearly show that this cognitive faculty has various components and that there are various steps through which infants acquire it (see Saxe et al. 2004). Brain imaging studies also tend to indicate the possibility of an involvement of many areas in this function (Blakemore and Decety 2001; Frith and Frith 2003; Gallagher and Frith 2003).

Given the complexity of the problem, it would be naive to claim that the mechanism described in the present study is *the* mechanism at the basis of mind read-

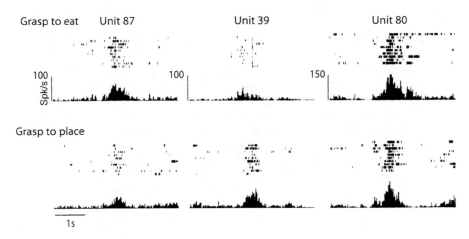

Fig. 2. Visual responses of IPL mirror neurons during the observation of grasping-to-eat and grasping-to-place done by an experimenter. Condtions as in Figure 1. (From Fogasso et al. 2005)

ing. Yet, the present data show for the first time a neural mechanism through which an important aspect of mind reading, understanding the intention of others, may be solved.

Emotion understanding

Up to now we have dealt with the neural mechanisms that enable individuals to understand "cold actions," that is, actions without any obvious emotional content. In social life, however, equally important, and maybe even more so, is the capacity to decipher emotions. Which mechanisms enable us to understand what others feel? Is there a mirror mechanism for emotions similar to that for cold action understanding?

It is reasonable to postulate that, as for action understanding, there are two basic mechanisms for emotion understanding that are conceptually different one from another. The first consists in cognitive elaboration of sensory aspects of others' emotional behaviors. The other consists in a direct mapping of sensory aspects of the observed emotional behavior on the motor structures that determine, in the observer, the experience of the observed emotion.

These two ways of recognizing emotions are experientially radically different. With the first, the observer understands the emotions expressed by others but does not feel them. He deduces them. A certain facial or body pattern means fear, another happiness, and that is it. No emotional involvement. Different is the case for sensory-motor mapping mechanism. In this case, the recognition occurs because the observed emotion triggers the feeling of the same emotion in the observing person. It is a first-person recognition. The emotion of the other penetrates the emotional life of the observer, evoking in him/her not only the observed emotion but also related emotional states and nuances of similar experiences.

As for cold action, our interest in this essay is the mechanisms underlying the direct sensory-motor mapping. For the sake of space, we will review data on one emotion only – disgust – for which rich empirical evidence has been recently acquired.

Disgust is a very basic emotion whose expression has an important survival values for the conspecifics. In its most basic, primitive form ("core disgust" Rozin et al. 2000) disgust indicates that something (e.g., food) that the individual tastes or smells is bad and, most likely, dangerous. Because of its strong communicative value, disgust is an ideal emotion for testing the direct mapping hypothesis.

Brain imaging studies showed that when an individual is exposed to disgusting odors or tastes, there is an intense activation of two structures: the amygdala and the insula (Augustine 1996; Royet et al. 2003; Small et al. 2003; Zald et al. 1998; Zald and Pardo 2000). The amygdala is a heterogeneous structure formed by several subnuclei. Functionally, these subnuclei form two major groups: the corticomedial group and the basolateral group. The former, phylogenetically more ancient, is related to the olfactory modality. It is likely that it is the signal increase in the corticomedial group that is responsible for the amygdala activation in response to disgusting stimuli.

Similarly to the amygdala, the insula is a heterogeneous structure. Anatomical connections revealed two main functional subdivisions in it an anterior "visceral" sector and a multimodal posterior sector (Mesulam and Mufson 1982). The anterior sector receives a rich input from olfactory and gustatory centers. In addition, the anterior insula receives an important input from the inferotemporal lobe, where, in the monkey, neurons have been found that respond to the sight of faces (Gross et al. 1972; Tanaka 1996). Recent data demonstrated that the insula is the main cortical target of interoceptive afferents (Craig 2002). Thus, the insula is not only the primary cortical area for chemical exteroception (e.g., taste and olfaction) but also for the interoceptive state of the body ("body state representation").

The insula is not an exclusively sensory area. In both monkeys and humans, electrical stimulation of the insula produces body movements (Kaada et al. 1949; Penfield and Faulk 1955; Frontera 1956; Showers and Lauer 1961; Krolak-Salmon et al. 2003). These movements, unlike those evoked by stimulation of classical motor areas, are typically accompanied by autonomic and viscero-motor responses.

Functional imaging studies in humans showed that, as in the monkey, the anterior insula receives, in addition to olfactory and gustatory stimuli, higher order visual information. Observation of disgusted facial expressions produces signal increase in the anterior insula. (Phillips et al. 1997, 1998; Sprengelmeyer et al. 1998; Schienle et al, 2002).

Recently, Wicker et al. (2003) carried out an fMRI study in which they tested whether the *same* insula sites that show signal increase during the experience of disgust also show signal increase during the observation of facial expressions of disgust.

The study consisted of olfactory and visual runs. In the olfactory runs, individuals inhaled disgusting and pleasant odorants. In the visual runs, the same participants viewed video-clips of individuals smelling a glass containing disgusting, pleasant and neutral odorants and expressing their emotions.

Disgusting odorants produced, as expected, a very strong signal increase in the amygdala and in the insula, with a right prevalence. In the amygdala, activation was also observed with pleasant odorants, with a clear overlap between the activations obtained with disgusting and pleasant odorants. In the insula, pleasant odorants produced a relatively weak activation located in a posterior part of the right insula; disgusting odorants activated the anterior sector bilaterally. The results of visual runs showed signal increases in various cortical and subcortical centers but not in the amygdala. The insula (anterior part, left side) was activated only during the observation of disgust.

The most important result of the study was the demonstration that precisely the same sector within the anterior insula that was activated by the exposure to disgusting odorants was also activated by the observation of disgust in others (Fig. 3). These data strongly suggest that the insula contains neural populations that become active both when the participants experience disgust and when they see it in others.

The notion that the insula mediates both recognition and experience of disgust is supported by clinical studies showing that, following lesions of the insula, patients have a severe deficit in understanding disgust expressed by others (Calder et al. 2000; Adolphs et al. 2003). This deficit is accompanied by blunted and reduced sensation of disgust. In addition, electrophysiological studies showed that

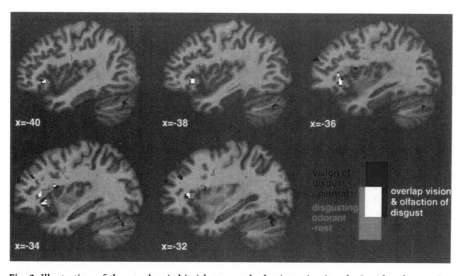

Fig. 3. Illustration of the overlap (white) between the brain activation during the observation (blue) and the feeling (red) of disgust. The olfactory and visual analyses were performed separately as random-effect analyses. The results are superimposed on parasagittal slices of a standard MNI brain.

sites in the anterior insula, whose electrical stimulation produced unpleasant sensations in the patient's mouth and throat, are activated by the observation of a face expressing disgust.

Taken together, these data strongly suggest that humans understand disgust, and most likely other emotions (see Carr et al. 2003; Singer et al., 2004), through a direct mapping mechanism. The observation of emotionally laden actions activates those structures that give a first-person experience of the same actions. By means of this activation, a bridge is created between others and us.

The hypothesis that we perceive emotion in others by activating the same emotion in ourselves has been advanced by various authors (e.g., Phillips et al. 1997; Adolphs 2003: Damasio 2003a; Calder et al. 2000; Carr et al. 2003; Goldman and Sripada 2003; Gallese et al. 2004). Particulary influential in this respect has been the work by Damasio and his coworkers (Adolphs et al. 2000; Damasio 2003a, b) According to these authors, the neural basis of empathy is the activation of an "as-if-loop," the core structure of which is the insula (Damasio 2003). These authors attributed a role in the "as-if-loop" also to somatosensory areas like SI and SII, conceiving the basis of empathy to be in the activation in the observer of those cortical areas where the body is represented.

Although this hypothesis is certainly possible, the crucial role of the insula, rather than of the primary somatosensory cortices, in emotion feeling strongly suggests that the neural substrate for emotions is not merely sensorial. It is more likely that the activation of the insula representation of the viscero-motor activity is responsible for the first-person feeling of disgust. As for the premotor cortex, it is plausible that in the insula there is a specific type of mirror neurons that match the visual aspect of disgust with its viscero-motor aspects. The activation of these (hypothetical) viscero-motor mirror neurons should underlie the first-person knowledge of what it means to be disgusted. The activation of these insular neurons should not necessarily produce the overt viscero-motor response. The overt response should depend on the strength of the stimuli and other factors. A neurophysiological study of insula neuron properties could be the direct test of this hypothesis.

Coda

The data reviewed in this essay show that the intuition of Adam Smith – that individuals are endowed with an altruistic mechanism that makes them share the "fortunes" of others – is strongly supported by neurophysiological data. When we observe others, we enact their actions inside ourselves and we share their emotions.

Can we deduce from this that the mirror mechanism is the mechanism from which altruistic behavior evolved? This is obviously a very hard question to answer. Yet, it is very plausible that the mirror mechanism played a fundamental role in the evolution of altruism. The mirror mechanism transforms what others do and feel in the observer's own experience. The disappearance of unhappiness in others means the disappearance of unhappiness in us and, conversely, the observation of happiness in others provides a similar feeling in ourselves. Thus, acting to render others happy – an altruistic behavior – is transformed into an egoistic behavior – we are happy.

Adam Smith postulated that the presence of this sharing mechanism renders the happiness of others "necessary" for human beings, "though he derives nothing from it except the pleasure of seeing it." This, however, appears to be a very optimist view. In fact, an empathic relationship between others and ourselves does not necessarily bring positive consequences to the others. The presence of an unhappy person may compel another individual to eliminate the unpleasant feeling determined by that presence, acting in a way that is not necessary the most pleasant for the unhappy person.

To use the mirror mechanism – a biological mechanism – strictly in a positive way, a further – cultural – addition is necessary. It can be summarized in the prescription: "Therefore all things whatsoever ye would that men should do to you, do ye even so to them: for this is the law and the prophets" (Matthew 7, 12) . This "golden rule," which is present in many cultures besides ours (see Changeux and Ricoeur 1998), uses the positive aspects of a basic biological mechanism inherent in all individuals to give ethical norms that eliminate the negative aspects that are also present in the same biological mechanism.

Acknowledgment

The study was supported by EU Contract QLG3-CT-2002–00746, Mirror, EU Contract IST-2000–29689, Artesimit, by Cofin 2004, and FIRB n. RBNE01SZB4. to G.R. and by EU Contract IST-004370, Robotclub, EU Cotract IST-001917, Neurobotics, and FIRB n. RBNE 018 ET9-004 to L.C.

References

Adolphs R (2003) Cognitive neuroscience of human social behaviour. Nature Rev Neurosci 4: 165–178.
Adolphs R, Damasio H, Tranel D, Cooper G, Damasio AR (2000) A role for somatosensory cortices in the visual recognition of emotion as revealed by three-dimensional lesion mapping. J Neurosci 20: 2683–2690.
Adolphs R, Tranel D, Damasio AR (2003) Dissociable neural systems for recognizing emotions. Brain Cogn 52: 61–69.
Allison T, Puce A, McCarthy G. (2000) Social perception from visual cues: role of the STS region. Trends Cogn Sci 4: 267–278.
Augustine JR (1996) Circuitry and functional aspects of the insular lobe in primates including humans. Brain Res Rev 22: 229–244.
Blakemore SJ, Decety J (2001) From the perception of action to the understanding of intention. Nature Rev Neurosci 2: 561.
Bruce C, Desimone R, Gross CG (1981) Visual properties of neurons in a polysensory area in superior temporal sulcus of the macaque. J Neurophysiol 46: 369–384.
Buccino G, Binkofski F, Fink GR, Fadiga L, Fogassi L, Gallese V, Seitz RJ, Zilles K, Rizzolatti G, Freund HJ (2001) Action observation activates premotor and parietal areas in a somatotopic manner: an fMRI study. Eur J Neurosci 13: 400–404.

Buccino G, Vogt S, Ritzl A, Fink GR, Zilles K, Freund HJ, Rizzolatti G (2004) Neural circuits underlying imitation of hand actions: an event related fMRI study. Neuron 42: 323–34.

Calder AJ, Keane J, Manes F, Antoun N, Young AW (2000) Impaired recognition and experience of disgust following brain injury. Nature Neurosci 3: 1077–1078.

Carey DP, Perrett DI, Oram MW (1997) Recognizing, understanding and reproducing actions. In: Jeannerod M, Grafman J (eds) Handbook of neuropsychology. Vol. 11: Action and cognition. Elsevier, Amsterdam.

Carr L, Iacoboni M, Dubeau MC, Mazziotta JC, Lenzi GL (2003) Neural mechanisms of empathy in humans: a relay from neural systems for imitation to limbic areas. Proc Natl Acad Sci USA 100: 5497–5502.

Changeux JP, Ricoeur P (1998) La nature et la règle. Odile Jacob, Paris.

Cochin S, Barthelemy C, Roux S, Martineau J (1999) Observation and execution of movement: similarities demonstrated by quantified electroencephalography. Eur J Neurosci 11: 1839–1842.

Craig AD (2002) How do you feel? Interoception: the sense of the physiological condition of the body. Nature Rev Neurosci 3: 655–666.

Damasio, A (2003a) Looking for Spinoza New York: Harcourt Inc.

Damasio A (2003b) Feeling of emotion and the self. Ann NY Acad Sci 1001: 253–261.

Di Pellegrino G, Fadiga L, Fogassi L, Gallese V, Rizzolatti G (1992) Understanding motor events: A neurophysiological study. Exp Brain Res 91: 176–80.

Fadiga L, Fogassi L, Pavesi G, Rizzolatti G (1995) Motor facilitation during action observation: a magnetic stimulation study. J Neurophysiol 73: 2608–2611.

Fogassi L, Gallese V, Fadiga L, Rizzolatti G (1998) Neurons responding to the sight of goal directed hand/arm actions in the parietal area PF (7b) of the macaque monkey. Soc Neurosci Abs 24:257.5.

Fogassi L, Ferrari PF, Gesierich B, Rossi S, Chersi f, Rizzolatti G (2005) Parietal Lobe: from action organisation to intention understanding. Science 308: 662–667.

Frith U, Frith CD (2003) Development and neurophysiology of mentalizing. Philos Trans R Soc Lond B Biol Sci 358: 459.

Frontera JG (1956) Some results obtained by electrical stimulation of the cortex of the island of Reil in the brain of the monkey (Macaca mulatta). J Comp Neurol 105: 365–394.

Gallagher HL, Frith CD (2003) Functional imaging of 'theory of mind'. Trends Cogn Sci 7: 77.

Gallese V, Fogassi L, Fadiga L, Rizzolatti G (2002) Action representation and the inferior parietal lobule. In: Prinz W, Hommel B (eds) Attention & Performance XIX. Common mechanisms in perception and action. Oxford University Press, Oxford.

Gallese V, Keysers C, Rizzolatti G (2004) A unifying view of the basis of social cognition. Trends Cogn Sci 8: 396–403.

Gangitano M, Mottaghy FM, Pascual-Leone A (2001) Phase specific modulation of cortical motor output during movement observation. NeuroReport 12: 1489–1492.

Gangitano M, Mottaghy FM, Pascual-Leone A (2004) Modulation of premotor mirror neuron activity during observation of unpredictable grasping movements. Eur J Neurosci 20: 2193–2202.

Goldman AI, Sripada CS (2004) Simulationist models of face-based emotion recognition. Cognition 94: 193–213.

Grafton ST, Arbib MA, Fadiga L, Rizzolatti G (1996) Localization of grasp representations in humans by PET: 2. Observation compared with imagination. Exp Brain Res 112: 103–111.

Grèzes J, Costes N, Decety J (1998) Top-down effect of strategy on the perception of human biological motion: a PET investigation. Cogn Neuropsychol 15: 553–582.

Grèzes J, Armony JL, Rowe J, Passingham RE (2003) Activations related to "mirror" and "canonical" neurones in the human brain: an fMRI study. Neuroimage 18: 928–937.

Gross CG, Rocha-Miranda CE, Bender DB (1972) Visual properties of neurons in the inferotemporal cortex of the macaque. J Neurophysiol 35: 96–111.

Hari R, Forss N, Avikainen S, Kirveskari S, Salenius S, Rizzolatti G (1998) Activation of human primary motor cortex during action observation: a neuromagnetic study. Proc. Natl Acad Sci USA 95: 15061–15065.

Iacoboni M, Woods RP, Brass M, Bekkering H, Mazziotta JC, Rizzolatti G (1999) Cortical mechanisms of human imitation. Science 286: 2526–2528.

Iacoboni M, Koski LM, Brass M, Bekkering H, Woods RP, Dubeau MC, Mazziotta JC, Rizzolatti G (2001) Reafferent copies of imitated actions in the right superior temporal cortex. Proc Natl Acad Sci USA 98: 13995–13999.

Jeannerod M (1988) The neural and behavioural organization of goal-directed movements. Clarendon Press, Oxford.

Johnson-Frey SH, Maloof FR, Newman-Norlund R, Farrer C, Inati S, Grafton ST (2003) Actions or hand-objects interactions? Human inferior frontal cortex and action observation. Neuron 39: 1053–1058.

Kaada BR, Pribram KH, Epstein JA (1949) Respiratory and vascular responses in monkeys from temporal pole, insula, orbital surface and cingulate gyrus: a preliminary report. J Neurophysiol 12: 347–356.

Kohler E, Keysers C, Umiltà MA, Fogassi L, Gallese V, Rizzolatti G (2002). Hearing sounds, understanding actions: action representation in mirror neurons. Science 297: 846–848.

Koski L, Wohlschlager A, Bekkering H, Woods RP, Dubeau MC (2002) Modulation of motor and premotor activity during imitation of target-directed actions. Cereb Cortex 12: 847–855.

Koski L, Iacoboni M, Dubeau MC, Woods RP, Mazziotta JC (2003) Modulation of cortical activity during different imitative behaviors. J Neurophysiol 89: 460–471.

Krolak-Salmon P, Henaff MA, Isnard J, Tallon-Baudry C, Guenot M, Vighetto A, Bertrand O, Mauguiere F (2003) An attention modulated response to disgust in human ventral anterior insula. Ann Neurol 53: 446–453.

Manthey S, Schubotz RI, von Cramon DY (2003). Premotor cortex in observing erroneous action: an fMRI study. Brain Res Cogn Brain Res 15: 296–307.

Mesulam MM, Mufson EJ (1982) Insula of the old world monkey. III: Efferent cortical output and comments on function. J Comp Neurol 212: 38–52.

Nishitani N, Hari R (2000) Temporal dynamics of cortical representation for action. Proc Natl Acad Sci USA 97: 913–918.

Nishitani N, Hari R (2002) Viewing lip forms: cortical dynamics. Neuron 36: 1211–1220.

Penfield W, Faulk ME (1955) The insula: further observations on its function. Brain 78: 445–470.

Perrett DI, Harries MH, Bevan R, Thomas S, Benson PJ, Mistlin AJ, Chitty AJ, Hietanen JK, Ortega JE (1989) Frameworks of analysis for the neural representation of animate objects and actions. J Exp Bio 146: 87–113.

Phillips ML, Young AW, Senior C, Brammer M, Andrew C, Calder AJ, Bullmore ET, Perrett DI, Rowland D, Williams SC, Gray JA, David AS (1997) A specific neural substrate for perceiving facial expressions of disgust. Nature 389: 495–498.

Phillips ML, Young AW, Scott SK, Calder AJ, Andrew C, Giampietro V, Williams SC, Bullmore ET, Brammer M, Gray JA (1998) Neural responses to facial and vocal expressions of fear and disgust. Proc R Soc Lond B Biol Sci 265: 1809–1817.

Puce A, Perrett D (2003) Electrophysiological and brain imaging of biological motion. Philosoph Trans Royal Soc Lond, Series B, 358: 435–445.

Rizzolatti G, Craighero L (2004) The mirror-neuron system. Annu Rev Neurosci 27: 169–192.

Rizzolatti G, Scandolara C, Matelli M, Gentilucci M (1981) Afferent properties of periarcuate neurons in macaque monkeys. I. Somatosensory responses. Behav Brain Res 2: 125–146.

Rizzolatti G, Fadiga L, Matelli M, Bettinardi V, Paulesu E, Perani D, Fazio F (1996) Localization of grasp representation in humans by PET: 1. Observation versus execution. Exp Brain Res 111: 246–252.

Rizzolatti G, Fogassi L, Gallese V (2001) Neurophysiological mechanisms underlying the understanding and imitation of action. Nature Rev Neurosci 2:661–670.

Royet JP, Plailly J, Delon-Martin C, Kareken DA, Segebarth C (2003) fMRI of emotional responses to odors: influence of hedonic valence and judgment, handedness, and gender. Neuroimage 20: 713–728.

Rozin R, Haidt J, McCauley CR (2000) Disgust. In: Lewis M, Haviland-Jones JM (eds) Handbook of Emotion. 2nd Edition. Guilford Press, New York, pp 637–653.

Saxe R, Carey S, Kanwisher N (2004) Understanding other minds: linking developmental psychology and functional neuroimaging. Annu Rev Psychol 55: 87–124.

Schienle A, Stark R, Walter B, Blecker C, Ott U, Kirsch P, Sammer G, Vaitl D (2002) The insula is not specifically involved in disgust processing: an fMRI study. Neuroreport 13: 2023–2026.

Showers MJC, Lauer EW (1961) Somatovisceral motor patterns in the insula. J Comp Neurol 117: 107–115.

Singer T, Seymour B, O'Doherty J, Kaube H, Dolan RJ, Frith CD (2004) Empathy for pain involves the affective but not the sensory components of pain. Science 303: 1157–1162.

Small DM, Gregory MD, Mak YE, Gitelman D, Mesulam MM, Parrish T (2003) Dissociation of neural representation of intensity and affective valuation in human gustation Neuron 39: 701–711.

Smith A (1759) The theory of moral sentiments (ed. 1976). Clarendon Press, Oxford.

Sprengelmeyer R, Rausch M, Eysel UT, Przuntek H (1998) Neural structures associated with recognition of facial expressions of basic emotions Proc R Soc Lond B Biol Sci 265: 1927–1931.

Strafella AP, Paus T (2000) Modulation of cortical excitability during action observation: a transcranial magnetic stimulation study. NeuroReport 11: 2289–2292.

Tanaka K (1996) Inferotemporal cortex and object vision. Ann Rev Neurosci. 19: 109–140.

Tomasello M, Call J (1997) Primate cognition. Oxford University Press, Oxford.

Tremblay C, Robert M, Pascual-Leone A, Lepore F, Nguyen DK, Carmant L, Bouthillier A, Theoret H (2004) Action observation and execution: intracranial recordings in a human subject. Neurology. 63: 937–938.

Umilta MA, Kohler E, Gallese V, Fogassi L, Fadiga L, Keysers C, Rizzolatti G (2001) "I know what you are doing": a neurophysiological study. Neuron 31: 155–165.

Visalberghi E, Fragaszy D. (2002). Do monkeys ape? Ten years after. In: Dautenhahn K, Nehaniv C (eds) Imitation in animals and artifacts. MIT Press, Boston. pp. 471–500

Wicker B, Keysers C, Plailly J, Royet JP, Gallese V, Rizzolatti G (2003) Both of us disgusted in my insula: the common neural basis of seeing and feeling disgust. Neuron 40: 655–664.

Yokochi H, Tanaka M, Kumashiro M, Iriki A (2003) Inferior parietal somatosensory neurons coding face-hand coordination in Japanese macaques. Somatosens Mot Res 20 : 115–125.

Zald DH, Pardo JV (2000) Functional neuroimaging of the olfactory system in humans. Int J Psychophysiol 36: 165–181.

Zald DH, Donndelinger MJ, Pardo JV (1998) Elucidating dynamic brain interactions with across-subjects correlational analyses of positron emission tomographic data: the functional connectivity of the amygdala and orbitofrontal cortex during olfactory tasks. J Cereb Blood Flow Metab 18: 896–905.

How does the brain know when it is right?

Wolf Singer[1]

This book is about the evolution of moral judgments. Obvious questions from a neurobiological perspective are, firstly, how the cognitive functions required for moral judgments have evolved and, secondly, to what extent the criteria for these judgments are innate or have been acquired through social learning during the individual's life time. It is of interest to distinguish between the criteria that have been acquired by trial and error during evolution and are expressed in the genetically determined functional architecture of the brain, and the criteria that are acquired by learning, i.e., by the experience-dependent shaping of the functional organization of the brain. In both cases, one would like to know which areas of the brain are engaged in moral judgments, where and how the respective criteria are stored and how this knowledge is read out for the preparation of decisions. The majority of contributions to this book address these important questions.

The purpose of this chapter is somewhat different, as it attempts to identify the neuronal constraints of judgments. The question is how the brain can actually know when the computational process that must precede decisions and judgments has converged towards a solution, and how the brain can evaluate the validity of the respective solution. Somehow we know from introspection, from our first-person perspective, when our deliberations have converged towards a result, how compatible this result is with our feelings – "whether it feels right" – and how reliable the conclusion is. We know whether there is a conflict between our rational deductions and our feelings and we know about the degree of uncertainty of our judgments. This evidence suggests to us that the brain can distinguish between processes that precede and lead towards results and the activity patterns that actually represent the result and, moreover, that the states representing the results are graded according to the degree of their reliability.

To the neurophysiologist who investigates brain functions with the registration of the electrical activity of nerve cells, the computational processes present themselves as highly dynamic and ever-changing activation patterns in which very large numbers of spatially distributed neurons participate. The activity of individual cells in turn consists of sequences of action potentials that vary in their frequency and in the precise temporal distribution of the individual discharges. Thus, any particular state is characterized by an exceedingly complex relational pattern, the features of which are defined by the topology of the participating cells

[1] Max Planck Institute for Brain Research, Deutschordenstr. 46, D-60528 Frankfurt/Main, Germany

Changeux et al.
Neurobiology of Human Values
© Springer-Verlag Berlin Heidelberg 2005

and the temporal relations among their discharges. The question is whether an activation pattern that represents a solution has a specific signature that allows the brain to distinguish such states from the activation patterns that occur during the computations that lead towards a result. If such a signature exists, it must be of a very general nature, because it must apply for all computational processes, irrespective of the modality in which they occur. Whether we have to resolve a perceptual ambiguity, a logical problem, or a moral conflict, we know when we have arrived at a solution and how reliable the solution is.

However, before entering the discussion about the possible nature of this signature, the possibility ought to be considered that our subjective experience of being able to distinguish between states of search and states of solutions is an illusion and that a search for such distinctions may be an ill-posed question. It is at least conceivable that our descriptions of neuronal processes that we derive from the third-person perspective of the scientific approach diverge from subjective experience, as is the case, for example, with the conflict between the subjective experience of free will and the deterministic neuronal processes that underlie decisions. Thus, one might argue that the distinction between computations and results is meaningless if the brain is a device that converts a continuous stream of input signals into a continuous stream of output activity or, in other terms, if the brain is just migrating continuously in a high-dimensional state space rather than moving from one fixed attractor to the next.

This analysis raises the question of whether there is more than just subjective evidence for a distinction between computations and solutions. The following arguments suggest that such a distinction is warranted. First, there is abundant psychophysical evidence that indicates that the time needed to reach perceptual decisions and to respond depends, to a crucial extent, on the difficulty of the task, suggesting that the underlying computational processes require a variable amount of time to converge towards a distinct solution and that input signals are not continuously converted into output signals under such conditions. Other evidence comes from the phenomena of perceptual ambiguity and rivalry. In these cases, there is no change in the input signals but brain states oscillate in an all-or-none manner between two stable solutions that, in this case, correspond to distinct perceptions. In the case of the Necker cube or the staircase paradigm, the brain alternates between two mutually exclusive interpretations of a visual scene. The same is the case with interocular rivalry. If two images are presented to the two eyes simultaneously and the images are so different that they cannot be fused into one image, perception switches between these two images in an all-or-none fashion, suggesting again that computational processes settle in discrete states. Similar conclusions can be drawn from studies on event-related potentials. These electrographical responses typically exhibit a succession of distinct peaks that allow one to trace the path of neuronal activity from primary sensory areas to motor cortex. Evidence suggests that these discrete peaks are not simply reflecting the migration of afferent volleys through the neuronal network. The wide spacing of the peaks, which can amount to several hundred milliseconds, excludes the possibility that the delays are a simple reflection of conduction delays. A more plausible and widely accepted interpretation is that these peaks reflect synchronized activity within the respective processing areas, whereby synchrony results either from internal interactions that converge towards brief episodes of highly synchronized activity or from a stimulus-locked resetting of ongoing oscillatory processes. These data suggest that computations within the respective areas converge at distinct times towards highly coherent states of activity that are then broadcast to the other processors.

More indirect arguments in favor of a distinction between computation and result states are derived from the constraints of learning processes. It is commonly held that the neuronal mechanism underlying the formation of both procedural and declarative memories consists of an activity-dependent, long-lasting modification of synaptic weights in large numbers of distributed neurons. It is obvious that such changes should occur only in instances when the conveyed activity is identified as appropriate in a broader behavioral context. Evidence is indeed available, in particular from developmental studies, that only those activation patterns are enabled to lead to long-lasting synaptic modifications that match certain criteria of adequacy (Singer 1995). The respective gating function is exerted by modulatory systems that originate in core structures of the brain whose activity is associated with arousal, attention, and reward (Singer 1990). These systems need to be informed when they should emit the permissive gating signals that allow synaptic gain changes to occur. Hence, a distinction between computational operations and results seems to be required to evaluate the behavioral significance of a result and to decide whether the corresponding activation pattern should lead to lasting changes in network properties. The activation dynamics of these reward systems are in line with this postulate. They reflect in amplitude and time course the adequacy of particular states in the context of sensory stimuli or motor acts (Fiorillo et al. 2003).

The putative signature of a result

The question of what the neuronal signature of a computational result could be is closely linked to the question of how cognitive objects and motor programs are represented in the brain. Two complementary coding strategies are considered at present. One is based on the notion that sensory signals are processed serially, whereby extraction and recombination of stimulus features alternate until cells at the top of the processing hierarchy are generated whose highly selective responses signal the presence of a particular sensory object (Barlow 1990). The counterpart of such smart neurons on the executive side are command neurons whose activation triggers a complex action pattern. In this conceptual framework, a result is the activation of a smart neuron. Because of their high selectivity, smart neurons are silent most of the time and become active only in the rare instances when afferent activity has succeeded in passing all the barriers set by the numerous filtering and thresholding operations in the chain of preceding processing stages. Likewise, the activity of a command neuron would signal that the preparation for a motor act has converged towards a concrete result that now requires execution. The rare firing of smart neurons would thus signal that all ambiguities have been resolved and that a computation has converged towards a solution.

A completely different signature for computational results is required if representations of cognitive objects and motor acts consist of the coordinated activation of very large numbers of neurons that may be distributed across different cortical areas, such as is postulated by concepts of assembly coding (Hebb 1949; for a detailed discussion of the differences between these coding strategies, see Singer 1999). In this case, the activity of individual cells cannot be interpreted as the result of a computation because, in assembly coding, neurons participate at different times in different assemblies, and their individual responses are only indicative of the presence of a particular subfeature of the content that is to be represented by an assembly. The activity of an individual neuron, when considered in isolation, does not indicate whether the neuron is actually part of an assembly and, if it is, in which assembly it is actually bound. In assembly coding, the result of a computational operation is the formation of an assembly and its stabilization for a period of time that is long enough to enable the assembly to affect the formation of other assemblies in sensory or executive structures. Donald Hebb, the most prominent initiator of concepts related to distributed processing and assembly coding, proposed that assemblies are a group of neurons that are more active than all the others because of cooperative reinforcement of their discharges through reciprocal excitatory connections. According to this proposal, the signature that identifies a neuron as part of an assembly would be its enhanced and sustained discharge rate. Accordingly, a mechanism that were to find out whether a particular assembly has stabilized would have to integrate over the activity of the neurons of a network and probe whether a certain constellation of neurons stands out because of particularly strong and sustained responses.

Due to the great success of research guided by the smart neuron concept, theoretical and experimental investigations of assemblies have experienced a renaissance only recently. In the course of these studies, it was noted that coordinated elevations in discharge rates alone are ambiguous signatures for assemblies for a variety of reasons. One reason is that the amplitude of a neuronal response depends on numerous variables that may be entirely unrelated to the formation of an assembly. Such variables are, for example, stimulus intensity, the appropriateness of a stimulus with respect to the receptive field properties of a neuron, or global changes in the excitability of neuronal networks. Another reason is the so-called superposition problem. If within a given neuronal network more than one assembly needs to be activated, which is the case when composite objects or spatially overlapping objects have to be represented, then it becomes difficult to find out which of the many neurons with enhanced responses actually belong to one assembly or the other (von der Malsburg 1999). Another problem arises with respect to processing speed. It has been pointed out that recognition and response times are often so short that decisions need to rely on only a few spikes per neuron (Fabre-Thorpe et al. 2001). For most of our cognitive and executive abilities, there is not enough time to integrate sequences of neuronal discharges over intervals long enough to permit the distinction of differences in discharge rate. Last but not least, the computational operations that lead to the formation of assemblies may go along with higher discharge rates than the stabilization of an assembly.

These shortcomings initiated the search for a signature of assemblies that is less ambiguous than elevated discharge rate and can be read out with the required speed. One proposal has been that this signature could be the precise temporal synchronization of the individual discharges of neurons participating in an assembly (von der Malsburg 1999; Singer 1999). The rationale behind this proposal is that rate enhancement and synchronization are equivalent from the point of view of receiving structures. Both processes raise the impact of activity patterns broadcast to target structures by exploiting either temporal summation (in the case of rate coding) or spatial summation (in the case of synchronization; Usrey and Reid 1999). In both cases, responses are selected for further joint processing by virtue of increasing the impact that they have in target structures. However, the difference is that, in the case of spike synchronization, the relations between the neuronal responses are defined with very high temporal precision. The reason is that the only discharges that profit from synchronization are those that are coincident within a few milliseconds with others, because of the short time constants of EPSPs in the dendrites of receiving cortical neurons (Ariav et al. 2003). As the synchronicity of discharges can be modified independently of discharge rate (Pipa and Grün 2003), the two options to increase the saliency of neuronal responses are orthogonal to each other. Discharge rates can be used to signal the presence of a particular feature and synchrony can be used to tag responses as related.

In conclusion, if precise synchronization of neuronal discharges is used as the signature of relatedness in assembly coding, assemblies can be formed and reconfigured at fast time scales, and if several assemblies have to be generated within the same neuronal matrix, they can be distinguished from one another with high temporal resolution (Singer and Gray 1995). If such a coding strategy were applied by the cerebral cortex, a computational result would be distinguished from the preceding computational operations by the fact that the neurons that have successfully organized themselves into an assembly emit precisely synchronized discharges over a limited period of time. A computational result would consist of the synchronous firing of a large number of distributed neurons. This firing would be a salient event because of its low probability and because of the strong impact that synchronous firing has in any target structure.

Evidence indicates that precise synchronization of spike discharges is frequently associated with an oscillatory modulation of neuronal activity (Gray and Singer 1989), especially when synchronization is established over larger distances and across different cortical areas (Engel et al. 1991 ; Roelfsema et al. 1997). Theoretical (Kopell et al. 2000) and experimental (König et al. 1995) analyses have demonstrated that such oscillatory patterning of neuronal responses greatly facilitates precise synchronization, and in vitro studies indicate that the network of inhibitory interneurons plays a crucial role in the generation of rhythmicity (Whittington et al. 1995). Following the discovery that neurons in the visual cortex exhibit oscillatory responses to light stimuli that can synchronize in a context-dependent way (Gray et al. 1989), numerous studies have been performed to search for the functional significance of neuronal synchrony. These studies have revealed that synchronous oscillatory activity is a ubiquitous phenomenon in the brain (for review, see Singer 2004). Different structures exhibit preferred oscillation frequencies but most cortical areas are capable of engaging in many different frequencies, whereby the preferred ranges depend on central states such as the different stages of sleep, arousal, and focused attention (Steriade et al. 1990). As a rule of thumb, oscillation frequency is inversely related to the spatial distance over which synchronicity is established. Short-range synchronization typically occurs in the high gamma ranges, whereas interareal long-distance synchronization in the awake brain is typically associated with oscillations in the beta-frequency range (Roelfsema et al. 1997; Brovelli et al. 2004). In agreement with the hypothesis that precise synchronization of discharges might serve as a universal tag of relatedness that serves to bind units temporarily into functionally coherent assemblies, synchronous oscillatory activity has been found to be associated with a large number of different cognitive and executive functions. These comprise early sensory processing in the olfactory, visual, auditory, and somatosensory modality, the preparation and execution of motor acts, the control of selective attention, and the formation of short- and long-term memories (for review, see Singer 1999; Tallon-Baudry and Bertrand 1999; Engel et al. 2001).

Synchronization of oscillatory activity as a signature of solutions

As mentioned above, interocular rivalry is one of the phenomena where perceptual processes have two equally likely solutions and where the respective neuronal networks settle in alternation in one of two stable states. Psychophysical evidence suggests that the activation patterns arriving from the two eyes compete for the first time within primary visual cortex (Lee and Blake 2004), and that it is the outcome of this competition that determines which activation pattern is propagated further and eventually gives rise to conscious experience. Multielectrode recordings performed in cats exposed to binocular rivalry have revealed that the activation patterns that have access to consciousness are those that succeed in engaging in precisely synchronized oscillations in the gamma frequency range (Fries et al. 1997, 2002). By analyzing the temporal coherence among the spiking patterns of distributed neurons, it was possible to predict which signals have access to consciousness and the control of eye movements. No such predictions were possible from measurements of the discharge rates of the same neurons. This finding suggests that the solution of a computational operation, in this case competition between the inputs from the two eyes, manifests itself in a state of coherent, precisely synchronized activity.

A similar result has been obtained with multielectrode recordings from the visual cortex of cats that had been exposed to so-called plaid patterns. These consist of two superimposed gratings that drift in orthogonal directions. Plaids give rise to two different, mutually exclusive percepts that alternate with each other. Subjects perceive either two independently moving gratings, one of which is transparent and slides on top of the other, or alternatively, two gratings that are fused and perceived as a single grid pattern that moves at slower speed in a direction that is intermediate between the respective motion directions of the component gratings. Which of the two perceptual solutions dominates could again be deduced from the analysis of the respective synchronization patterns and not from measurements of the discharge rates of neurons in early visual areas. In conditions where the animals perceived component motion, neurons split into two groups, whereby the discharges of cells responding to the same grating but not those responding to different gratings synchronized. Thus, in this case, the solution of the perceptual ambiguity consisted of the formation of two different assemblies of cells whose respective activities were coherent. The activity of the very same cells converged towards a different coherent state when conditions were such that the animal perceived the two gratings as fused into a grid that moved coherently in the intermediate direction. In this case, all cells, irrespective of whether they are driven by either or other grating, synchronize their responses, and thus form one large assembly (Castelo-Branco et al. 2000). Thus, the alternative solution consists again of a state in which a group of neurons, in this case in a different constellation, synchronizes its discharges. Within the range of temporal resolution that can be obtained with cross-correlation analysis, the transitions between these two solutions are abrupt, as is the switch between the two perceptual alternatives.

In processes that involve the cooperation of different cortical areas, which is the case in virtually all cognitive and executive functions, a solution must consist of the successful coordination of the activity in various cortical areas. Several lines of evidence suggest that the brain does indeed strive towards such states of large-scale coherence. When animals are trained to perform a sensory discrimination task and to respond within a limited period of time, they can anticipate that, following the alerting signal, they will have to engage a number of cortical areas to accomplish the task. If the task involves a visual discrimination, the various areas of the visual cortex – the polymodal parietal areas required for the preparation of targeting movements and the respective premotor and motor areas – have to be engaged. Electrophysiological evidence indicates that these areas engage in synchronous oscillatory activity in the beta-frequency range immediately after the animal has been alerted by the warning signal that the next discrimination trial will start. When the stimulus that needs to be discriminated is presented, this synchronization increases and becomes even more precise, but once the task is accomplished, this highly coordinated pattern of synchronous oscillations collapses and gives way to an unsystematic waxing and waning of idling, low frequency oscillations (Roelfsema et al. 1997). Again, this result suggests that the brain strives for coherent states when it needs to arrive at solutions that can be translated into motor acts.

Similar conclusions can be drawn from EEG and MEG studies in human subjects. Here, the resolution of perceptual problems such as figure-ground distinctions, the binding of distributed features into perceptual object and the formation of polymodal associations are associated with the appearance of synchronized oscillatory activity over the respective cortical areas and the establishment of precise phase synchronization (for review, see Tallon-Baudry and Bertrand 1999; Singer 2004). These episodes of coordinated activity are of short duration, the time course corresponding closely to the time when subjects become aware of the results of the respective computational operations. These results are all compatible with the hypothesis that the results of computational operations consist of transient states that are characterized by a high degree of synchronous firing in populations of neurons that, depending on the complexity of the respective computations, are distributed more or less widely across the cortical mantle.

Given the statistics of neuronal spike trains, such highly synchronized states, even if they are only very brief and comprise in extremis only a single spike per neuron, are exceedingly improbable. While pairwise coincidences are rather likely events, the coincident firing of hundreds or thousands of distributed cells is an extremely unlikely event and hence is of high significance. It could thus serve as an ideal signature to distinguish computational operations that strive towards states of coherence from the coherent state itself. As there is no observer in the brain, such coherent states would be the not-further-reducible result of a computation and the substrate of phenomenal awareness (Singer 2005). The reward systems could read such states if their projection neurons were tuned to the precisely synchronized input that they would have to sample from specialized neurons that are distributed across the cortical mantle, and that are themselves coincidence-sensitive and become active only if a sufficient number of cells in their environment engages in synchronized activity. The reliability of a solution could then be derived from variables that indicate the robustness of such states of synchrony, such as the precision and the duration of synchrony and the number of neurons and subsystems participating in such states. With appropriate thresholding, the neurons of the internal evaluation or reward systems could express the reliability of a computational result by the vigor of their discharges. The secondary consequences of the activation of these systems could be the "feeling" of how reliable a computational result is. The preparation of a motor response, which is bound to follow such high-impact states whether it remains covert or is actually executed, may then be the event that makes us aware of the fact that we do have a solution.

The detectability of solutions with non-invasive techniques

In case the signature of a solution is the coherence of neuronal activation patterns, the electrophysiological variables that can be assessed with the non-invasive techniques of EEG and MEG recordings are ideally suited to detect the instances at which solutions are achieved. Per definition, these techniques can only pick up signals that result from highly synchronized activity, because neuronal activity that lacks temporal coherence adds up to zero when assessed with techniques that average over thousands of neurons. The amplitude of EEG and MEG signals is directly proportional to the number of synchronously active neurons and to the degree of precision with which these neurons are synchronized. Hence, EEG and MEG recordings are particularly well-suited to assess results of computational operations, provided that some of the above speculations are valid.

Quite unexpectedly, recent investigations into the correlation between neuronal activity and hemodynamic responses suggest that the BOLD signal that serves as the basis of fMRI investigations does not only depend on the discharge rate of neurons but, to a crucial extent, also on the coherence of neuronal firing (Logothetis et al. 2001). A recent experiment in which optical recording of the hemodynamic response was combined with direct recordings of neuronal activity in the visual cortex of cats revealed that the amplitudes of the hemodynamic responses varied over a wide range, even if stimulus conditions were held strictly constant. The same holds for the variability of the neuronal responses. The amplitude of the responses of individual cells is also highly variable, as is the engagement in synchronous oscillatory firing. Surprisingly, trial by trial analysis revealed only a weak correlation between the actual discharge rate of neuronal population responses and the corresponding hemodynamic signal. A remarkably good correlation existed, however, between the amplitude of the hemodynamic response and the engagement of neurons in synchronous oscillatory firing in the high frequency range (Niessing et al. 2005). A likely explanation for this unexpected finding is that hemodynamic responses do not only depend on the firing of pyramidal cells that are commonly recorded in electrophysiological experiments but also and probably to a larger extent on the activity of inhibitory interneurons. When a cortical network engages in high-frequency gamma oscillations, the network of inhibitory interneurons becomes highly active and discharges essentially without cycle skipping in phase with the respective oscillations. This frequently leads to a reduction of pyramidal cell firing but, at the same time, increases the synchronicity of the responses of pyramidal cells. As inhibitory interneurons distribute most of their synapses in their immediate vicinity, and as the metabolically expensive processes are not the generation of action potentials but the molecular events associated with transmitter release and calcium entry, it could well be that the network of active inhibitory interneurons contributes more to the hemodynamic response than the firing of pyramidal cells, which distribute their synapses over large distances. It is, thus, conceivable – and this might be good news for the ever-growing community of researchers exploiting fMRI as a tool to assess brain activity – that the BOLD signal is particularly sensitive to states that have the signature of a computational result. Support for the hypotheses proposed above comes from a recent study on neuronal correlates of the "Eureka" phenomenon, using verbal problems whose solutions lead to sudden insight. The authors found that such Eureka experiences are associated with an enhanced BOLD response and a brief burst of gamma oscillations in the anterior part of the superior temporal gyrus of the right hemisphere (Jung-Beeman et al. 2004).

Concluding Remarks

The feeling of having reached the right decision might, thus, be associated with brain states that are distinguished by particularly high coherence. As is hypothesized for the integration of multimodal sensory signals into coherent percepts, such states are likely to emerge when the distributed processes occurring simultaneously in different cortical areas can be brought to match. In the case of moral decisions, such could be the case when agreement is reached among the "votes" of the various centers that contribute different aspects, such as innate incentives, early imprinted convictions and later acquired rational arguments.

References

Ariav G, Polsky A, Schiller J (2003) Submillisecond precision of the input-output transformation function mediated by fast sodium dendritic spikes in basal dendrites of CA1 pyramidal neurons. J Neurosci 23: 7750-7758

Barlow HB (1990) The mechanical mind. Annu Rev Neurosci 13: 15-24

Brovelli A, Ding M, Ledberg A, Chen Y, Nakamura R, Bressler SL (2004) Beta oscillations in a large-scale sensorimotor cortical network: Directional influences revealed by Granger causality. Proc Natl Acad Sci USA 101: 9849-9854

Castelo-Branco M, Goebel R, Neuenschwander S, Singer W (2000) Neural synchrony correlates with surface segregation rules. Nature 405: 685-689

Engel AK, König P, Kreiter AK, Singer W (1991) Interhemispheric synchronization of oscillatory neuronal responses in cat visual cortex. Science 252: 1177-1179

Engel AK, Fries P, Singer W (2001) Dynamic predictions: oscillations and synchrony in top-down processing. Nat Rev Neurosci 2: 704-716

Fabre-Thorpe M, Delorme A, Marlot C, Thorpe S (2001) A limit to the speed of processing in ultra-rapid visual categorization of novel natural scenes. J Cogn Neurosci 13: 171-180

Fiorillo CD, Tobler PN, Schultz W (2003) Discrete coding of reward probability and uncertainty by dopamine neurons. Science 299: 1898-1902

Fries P, Roelfsema PR, Engel AK, König P, Singer W (1997) Synchronization of oscillatory responses in visual cortex correlates with perception in interocular rivalry. Proc Natl Acad Sci USA 94: 12699-12704

Fries P, Schröder J-H, Roelfsema PR, Singer W, Engel AK (2002) Oscillatory neuronal synchronization in primary visual cortex as a correlate of stimulus selection. J Neurosci 22: 3739-3754

Gray CM, Singer W (1989) Stimulus-specific neuronal oscillations in orientation columns of cat visual cortex. Proc Natl Acad Sci USA 86: 1698-1702

Gray CM, König P, Engel AK, Singer W (1989) Oscillatory responses in cat visual cortex exhibit inter-columnar synchronization which reflects global stimulus properties. Nature 338: 334-337

Hebb DO (1949) The organization of behavior. John Wiley & Sons, New York

Jung-Beeman M, Bowden EM, Haberman J, Frymiare JL, Arambel-Liu S, Greenblatt R, Reber PJ, Kounios J (2004) Neural activity when people solve verbal problems with insight. Publ Lib Sci (PLOS) Biol 2: 0500-0510

König P, Engel AK, Singer W (1995) Relation between oscillatory activity and long-range synchronization in cat visual cortex. Proc Natl Acad Sci USA 92: 290-294

Kopell N, Ermentrout GB, Whittington MA, Traub RD (2000) Gamma rhythms and beta rhythms have different synchronization properties. Proc Natl Acad Sci USA 97: 1867-1872

Lee S-H, Blake R (2004) A fresh look at interocular grouping during binocular rivalry. Vis Res 44: 983-991

Niessing J, Ebisch B, Schmidt KE, Niessing M, Singer W, Galuske RAW (2005) Hemodynamic signals correlate tightly with synchronized gamma oscillations but only loosely with spiking activity. Science, 309: 948-951

Logothetis NK, Pauls J, Augath M, Trinath T, Oeltermann A (2001) Neurophysiological investigation of the basis of the fMRI signal. Nature 412: 150-157

Pipa G, Grün S (2003) Non-parametric significance estimation of joint-spike events by shuffling and resampling. Neurocomputing 52-54: 31-37

Roelfsema PR, Engel AK, König P, Singer W (1997) Visuomotor integration is associated with zero time-lag synchronization among cortical areas. Nature 385: 157-161

Singer W (1990) The formation of cooperative cell assemblies in the visual cortex. J Exp Biol 153: 177-197

Singer W (1995) Development and plasticity of cortical processing architectures. Science 270: 758-764

Singer W (1999) Neuronal synchrony: A versatile code for the definition of relations? Neuron 24: 49–65

Singer W (2004) Synchrony, oscillations, and relational codes. In: Chalupa LM, Werner JS (eds) The visual neurosciences. Cambridge: The MIT Press, A Bradford Book, pp 1665–1681

Singer W (2005) Large-scale temporal coordination of cortical activity as prerequisite for conscious experience. In: Nirit S (ed) Companion to consciousness. Malden, MA: Blackwell Publishing Inc. , in press

Singer W, Gray CM (1995) Visual feature integration and the temporal correlation hypothesis. Annu Rev Neurosci 18: 555–586

Steriade M, Gloor P, Llinas RR, Lopes da Silva FH, Mesulam MM (1990) Basic mechanisms of cerebral rhythmic activities. Electroenceph Clin Neurophysiol 76: 481–508

Tallon-Baudry C, Bertrand O (1999) Oscillatory gamma activity in humans and its role in object representation. Trends Cogn Sci 3: 151–162

Usrey WM, Reid RC (1999) Synchronous activity in the visual system. Annu Rev Physiol 61: 435–456

von der Malsburg C (1999) The what and why of binding: the modeler's perspective. Neuron 24: 95–104

Whittington MA, Traub RD, Jefferys JGR (1995) Synchronized oscillations in interneuron networks driven by metabotropic glutamate receptor activation. Nature 373: 612–615

Cerebral basis of human errors

Olivier Houdé[1]

> *Our method correctly explains how we do not fall into error*
> *and how deductions are to be discovered so that we reach the knowledge*
> *of everything"*
> (René Descartes, 1628)

Summary

As stated by Jean-Pierre Changeux in his last book, "The Physiology of Truth," objective knowledge does exist, and our brains are naturally equipped to recognize it. The results presented here provide the first insights into 1) the cerebral basis of reasoning errors, and 2) the neurocognitive dynamics that lead the human brain toward logical truth.

Introduction

Ever since Aristotle, it has been known that the essence of the human mind is the "logos." It encompasses both reason (i.e., logic) and language. However, as the 17th century French philosopher René Descartes demonstrated, an important challenge for humans is the implementation of deductive rules for redirecting the mind from reasoning errors to logical thinking, which Descartes called the "Method."

Current research on the cognitive psychology of deduction has confirmed that most individuals do not spontaneously apply the principles of logic in problem solving; their reasoning is often biased by misleading strategies (Evans 1989, 1998, 2003). Research on judgment and decision-making – initially presented in a series of fundamental works by Amos Tversky and Daniel Kahneman, the 2002 Nobel Prize Laureate in Economics – emphasized the role of short-cut heuristics in probability judgment and the cognitive biases that resulted from them (Kahneman et al. 1982; Kahneman and Tversky 2000). However, as stated by Jean-Pierre Changeux in his last book, "The Physiology of Truth," objective knowledge does exist, and our brains are naturally equipped to recognize it (Changeux 2004). Thus, one of the crucial challenges for cognitive and educational neuroscience

---

[1] Université Paris-5 et Centre Cyceron, Caen, CNRS et CEA, France; e-mail: olivier.houde@paris5.sorbonne.fr

Changeux et al.
Neurobiology of Human Values
© Springer-Verlag Berlin Heidelberg 2005

today is to discover the brain mechanisms that enable us to shift from reasoning errors to logical thinking.

Neuroimaging data on this topic are scarce. In our research, conducted with the collaboration of Bernard and Nathalie Mazoyer at the brain imaging centre, Cyceron, in Caen, France, we investigated three questions: 1) Do we reason with logic? 2) Why do we make reasoning errors (rather than reasoning according to the logical truth table)? 3) Can emotions help us reason? (Houdé and Tzourio-Mazoyer 2003). This last question is in line with Antonio and Hanna Damasio's work on emotion and reasoning (Damasio 1994, 1999, 2003).

Do we reason with logic?

Contrary to Jean Piaget's theory (Inhelder and Piaget 1964; Piaget 1984), which described a logical stage of thinking as of the age of 14 or 15, new studies on cognitive psychology of reasoning have shown that adolescents and adults consistently make deduction errors in certain tasks, due to what are called "reasoning biases" (Evans 1989, 1998, 2003; Houdé 2000; Houdé and Moutier 2004). For example, when the task was given to first read a rule such as, "If there is not a red square on the left, then there is a yellow circle on the right," and then to select two geometrical shapes that would make the rule *false*, most subjects spontaneously placed a red square on the left of a yellow circle, believing they were completing the task correctly. This logic error is caused by what Jonathan Evans has called the "matching bias" (Evans 1998). Subjects usually respond by using the shapes that are mentioned in the rule rather than reasoning according to the logical truth table, which, if used, would lead them to choose a case where the antecedent of the rule is true (i.e., not a red square) and the consequent is false (i.e., not a yellow circle) such as, a blue square to the left of a green diamond. The logical response, therefore, requires the subjects to resist the elements perceived in the rule: that is, to inhibit the matching bias (Houdé 2000). This is a good example of high-order abstraction, where logic needs to resist perception.

Why do we make reasoning errors?

In an attempt to understand the type of error shown in the above example, we hypothesized that adolescents and adults have two competing reasoning strategies in their "neural/mental work space" (Changeux 2004; Dehaene et al. 1998) – one logical, the other perceptual – and that they have trouble inhibiting the perceptual one. Following this analysis, the difficulty lies not in mental logic per se but in executive function, in this case, inhibition. This has also been shown in infants and children tested for elementary cognitive acquisitions (Diamond 1991; Houdé 2000).

To demonstrate this analysis, we conducted experimental psychology studies that tested the effectiveness of a "de-biasing (or error-correction) paradigm."

These studies were based on two experimental training conditions using the same type of deductive logic task but with different materials: 1) training in inhibition of the perceptual strategy (matching bias), and 2) training in logic only. These conditions were compared to a control condition of simple task repetition using the same design without training. We found that only inhibition training proved effective in reducing errors. We interpreted this to mean that an executive blocking mechanism was indeed what these adolescents and adults were lacking, not logic or practice (Houdé 2000), even though in some cases logical training is useful.

Following this initial testing, we carried out a brain imaging study to observe the cerebral changes that occurred before and after training under the matching bias-inhibition condition (Houdé et al. 2001). We found a clear shift in cerebral activity from the posterior part of the brain (or the "perceptual brain") before training to the prefrontal part after training, that is, at the moment when the error-to-logic shift occurred. According to Joaquin Fuster's interpretation of our results in his latest book, "Cortex and Mind" (Fuster 2003), "the exercise of logical reasoning seems to overcome [or to inhibit] the biasing influences from posterior cortex and to lend to prefrontal cortex the effective control of the reasoning task" (p. 231).

So, thanks to neuroimaging, it is now possible to know what happens in the human brain when a logic error is made. We can thus assert that there is a biological reality behind irrationality, as the philosopher Stephen Stich hypothesized in his book, "The Fragmentation of Reason" (Stich 1990), or, if we wish to avoid speaking of irrationality, we can say, as Evans (2003) stated, that there are "two minds [or two rationalities] in one brain" (p. 458).

Can emotions help us reason?

Contrary to Descartes' well-known dichotomy between reason and emotion, Damasio offers eloquent support for the view that "the good use of reason" depends on emotion (Bechara et al. 1997; Damasio 1994, 1999, 2003). In studying Phineas Gage's lesion (Damasio et al. 1994; Harlow 1848) and other more recent cases, Hanna and Antonio Damasio have shown that ventromedial prefrontal damage causes defects in reasoning/decision making, emotion, and self-feeling. In line with their contributions, we hypothesized that there may be a close tie between emotion, self-feeling, and reasoning error inhibition in the human brain (Houdé et al. 2001, 2003; Houdé and Tzourio-Mazoyer 2003).

In our previous study (Houdé et al. 2000), we noted that training under the matching bias-inhibition condition incorporated emotional warnings of the error risk that were not present under the logic-only training condition. We then compared the impact of these two training conditions and found that, under the matching bias-inhibition condition (at the moment when the error-to-logic shift occurred), greater cerebral activity was observed in the right ventromedial prefrontal cortex (Houdé et al. 2001, 2003), that is, neural activity was present in the location of the lesion in Gage and in Damasio's patients. These data suggest that in healthy subjects this paralimbic area (Mesulam 2000) participates in getting the mind intuitively on the "logical track".

The right ventromedial prefrontal cortex could, therefore, be the emotional component (internal warning/self-feeling) of the brain's error-correction device. More precisely, together with the anterior cingulated cortex, it could correspond to the brain area that detects the conditions under which logical reasoning errors might occur (Botvinick et al. 2004; Bush et al. 2000; Carter et al. 1998; Houdé 2003; Houdé and Tzourio-Mazoyer 2003; MacDonald et al. 2000). Other brain imaging studies have stressed the role of the medial part of the prefrontal cortex in the emotional evaluation of error risks in domains related to logical cognition, notably, in the rapid processing of monetary gains and losses during economic reasoning (Gehring and Willoughby 2002).

We also know that the medial part of the prefrontal cortex is involved in moral cognition (Casebeer 2003), and that early damage in this region in infancy causes impairment of moral behavior and moral knowledge during social development (Anderson et al. 1999, from Damasio's group). On the possible tie between moral cognition and logic, one should recall that the child psychologist Jean Piaget had remarkable insight when he stated that "logic should be the moral of cognition as moral is the logic of action" (see Vidal 1994).

In conclusion, our results provide the first insights into 1) the cerebral basis of reasoning errors and 2) the neurocognitive dynamics that lead the human brain toward logical truth. Along with other results on mathematical cognition (Dehaene 1997; Dehaene et al. 1999; Houdé and Tzourio-Mazoyer 2003), our discoveries argue for the neurobiology of truth (Changeux 2004) and of human values.

References

Anderson S, Bechara A, Damasio H, Tranel D, Damasio A (1999) Impairment of social and moral behavior related to early damage in human prefrontal cortex. Nature Neurosci 2: 1032–1037

Bechara A, Damasio H, Tranel D, Damasio, A (1997) Deciding advantageously before knowing the advantageous strategy. Science 275: 1293–1295

Botvinick M, Cohen J, Carter C (2004) Conflict monitoring and anterior cingulate cortex: an update. Trends Cogn Sci 8: 540–546

Bush G, Luu P, Posner M (2000) Cognitive and emotional influences in anterior cingulate cortex. Trends Cogn Sci 4: 215–222

Carter C, Braver T, Barch D, Botvinick M, Douglas N, Cohen J (1998) Anterior cingulate cortex, error detection, and the online monitoring of performance. Science 280: 747–749

Casebeer W (2003) Moral cognition and its neural constituents. Nature Rev Neurosci 4: 840–847

Changeux J-P (2004) The physiology of truth. Cambridge, MA: Harvard University Press

Damasio A (1994) Descartes' error. New York: Putnam

Damasio A (1999) The feeling of what happens. New York: Harcourt

Damasio A (2003) Looking for Spinoza. New York: Harcourt

Damasio H, Grabowski T, Frank R, Galaburda A, Damasio A (1994) The return of Phineas Gage: clues about the brain from the skull of a famous patient. Science 264: 1102–1105

Dehaene S (1997) The number sense. New York: Oxford University Press

Dehaene S, Kerszberg M, Changeux J-P (1998) A neuronal model of a global workspace in effortful cognitive tasks. Proc Natl Acad Sci USA, 95: 14529–14534

Dehaene S, Spelke E, Pinel P, Stanescu R, Tsivkin S (1999) Sources of mathematical thinking: Behavioral and brain-imaging evidence. Science 284: 970–974

Descartes R (1628/1961) Rules for the direction of the mind. Indianapolis: Bobbs-Merrill

Diamond A (1991) Neuropsychological insights into the meaning of object concept development. In: Carey S, Gelman R (eds), The epigenesis of mind. Hillsdale: Erlbaum, pp. 67–110

Evans J (1989) Bias in human reasoning. London: Erlbaum

Evans J (1998) Matching bias in conditional reasoning. Thinking Reasoning 4: 45–82

Evans J (2003) In two minds: dual-process accounts of reasoning. Trends Cogn Sci 7: 454–459

Fuster J (2003) Cortex and mind. New York: Oxford University Press

Gehring W, Willoughby A (2002) The medial frontal cortex and the rapid processing of monetary gains and losses. Science 295: 2279–2282

Harlow J (1848) Passage of an iron rod through the head. Boston Med Surg J 39: 389–393

Houdé O (2000) Inhibition and cognitive development: object, number, categorization, and reasoning. Cogn Dev 15: 63–73

Houdé O (2003) Consciousness and unconsciousness of logical reasoning errors in the human brain. Behav Brain Sci 25: 341

Houdé O, Moutier S (2004). Reasoning and rationality. In: Houdé O (ed) Dictionary of cognitive science. New York: Routledge/Taylor & Francis (Psychology Press), pp. 310–312

Houdé O, Tzourio-Mazoyer N (2003) Neural foundations of logical and mathematical cognition. Nature Rev Neurosci 4: 507–514

Houdé O, Zago L, Mellet E, Moutier S, Pineau A, Mazoyer B, Tzourio-Mazoyer N (2000) Shifting from the perceptual brain to the logical brain: The neural impact of cognitive inhibition training. J Cogn Neurosci 12: 721–728

Houdé O, Zago L, Crivello F, Moutier S, Pineau A, Mazoyer B, Tzourio-Mazoyer N (2001) Access to deductive logic depends on a right ventromedial prefrontal area devoted to emotion and feeling: Evidence from a training paradigm. NeuroImage 14 : 1486–1492

Houdé O, Zago, L, Moutier S, Crivello F, Mellet E, Pineau A, Mazoyer B, Tzourio-Mazoyer N (2003) Can emotions help us reason? Brain Cogn 51: 233–234

Inhelder B, Piaget J (1964) The growth of logical thinking. New York: Basic Books

Kahneman D, Tversky A (2000) Choices, values, and frames. New York: Cambridge University Press

Kahneman D, Slovic P, Tversky A (1982) Judgment under uncertainty: heuristics and biases. New York: Cambridge University Press

MacDonald A, Cohen J, Stenger V, Carter C (2000) Dissociating the role of the dorsolateral prefrontal and anterior cingulate cortex in cognitive control. Science 288: 1835–1838

Mesulam M (2000) Principles of behavioral and cognitive neurology. New York: Oxford University Press

Piaget J (1984) Piaget's theory. In: Mussen PH (ed), Handbook of child psychology. Vol. 1. New York: Wiley

Stich, S. (1990). The fragmentation of reason. Cambridge: The MIT Press, pp. 103–128

Vidal F (1994) Piaget before Piaget. Cambridge: Harvard University Press.

How a Primate Brain Comes to Know Some Mathematical Truths

Stanislas Dehaene[1]

Summary

What are the origins of mathematical truth? How can a finite and fallible human brain, created through the approximate tinkering of evolution rather than by an omnipotent designer, come to possess universal mathematical knowledge? Furthermore, "how is it possible that mathematics, a product of human thought that is independent of experience, fits so excellently the objects of physical reality?" (Einstein). Without pretending to address directly those difficult philosophical challenges, my research aims at understanding the cerebral origins of one of the foundations of mathematics, the concept of number. Both the human and the macaque parietal lobes contain neural representations of approximate number. Behavioral studies in the Mundurukú, an Amazonian people with very few number words, indicate that the concept of approximate number is independent of language and universally shared, but that exact calculation is not. These data lead to a mixed scenario for the origins of arithmetic. Evolution provides us with an approximate number sense, shared with our primate cousins. This representation then serves as the foundation for cultural constructions such as words, digits, and counting, which extend those abilities into the realm of exact arithmetic.

Introduction: the puzzles of mathematical truth

Disagreement over what constitutes truth is nowhere as evident as in the domain of mathematics. Many mathematicians believe that mathematics is a domain of pure truths that predate the human mind and have an abstract existence, independent of the mediocre human ability to discover them. Hardy, for instance, stated: "I believe that mathematical reality lies outside us, that our function is to discover or *observe* it, and that the theorems which we prove, and which we describe grandiloquently as our "creations," are simply our notes of our observations."

A similar statement was made by the French mathematician, Charles Hermite:

"I believe that the numbers and functions of analysis are not the arbitrary product of our spirits; I believe that they exist outside of us with the same

---

[1] INSERM Unit 562 "Cognitive Neuroimaging", Service Hospitalier Frédéric Joliot, CEA, 4, Place du Général Leclerc, 91401 Orsay, France; e-mail: dehaene@shfj.cea.fr

character of necessity as the objects of objective reality; and we find or discover them and study them as do the physicists, chemists, and zoologists."

For the modern neurobiologist, however, the Platonist position seems hard to defend – as unacceptable, in fact, as Cartesian dualism is as a scientific theory of the brain. What is this mysterious stuff of which mathematical reality is made of? What else could it be, but the creation of complex, interlocking assemblies of neurons within our brains? In his dialog with the French mathematician Alain Connes, the neurobiologist Jean-Pierre Changeux makes a statement which is at once bold, and yet obvious to any neuroscientist: "Mathematical objects correspond to physical states of our brain" (Changeux and Connes 1989).

The neurobiologist's "realist" stance, however, faces two further challenges that philosophers of mathematics have repeatedly raised. First, how can a finite and fallible human brain, created through the approximate tinkering of evolution rather than by an omnipotent designer, come to know some absolute mathematical truths? Whether we perceive faces, understand speech, or perform complex motor acts, the performance of our brains is usually limited, imperfectly reproducible and prone to error. What, then, is different about mathematical activity, that we seem to be able to freely wander in infinite spaces and to universally agree on facts such as $1+1=2$, or $e^{i\pi} = -1$?

A second challenge concerns what Eugene Wigner has called the "unreasonable effectiveness" of mathematics as it applies to the natural world. Albert Einstein summarized the problem succinctly: "How is it possible that mathematics, a product of human thought that is independent of experience, fits so excellently the objects of physical reality?" An example of this amazing dialog back and forth between mathematics and physics is provided by Pascal's triangle, a mathematical object that was independently discovered many times, including by Chinese mathematicians. Although the triangle is generated by simple serial additions, the base of the triangle also gives the binomial distribution, which converges to the Gaussian curve – a pure mathematical object, yet one that provides an amazingly tight fit to many natural phenomena, such as the microscopic movement of particles (Brownian motion).

A biological approach to mathematical truth

Without pretending to resolve these classical problems in the philosophy of mathematics, I would like to suggest that a biological "selectionist" approach, based upon the notion of evolved "prerepresentations" and their mental selection (Changeux 1983, 2002; Changeux and Dehaene 1989), may turn them into an addressable problem. During its evolution, our primate brain has been endowed with elementary representations that are adequate to certain aspects of our environment. These internalized representations of time, space, and number provide the foundations of mathematics. Further below, I will provide evidence that the concept of number, for instance, has a long evolutionary history and is spontaneously available to many animal species, to human infants, and to uneducated human adults. Thus, it should not be surprising that we all agree on the importance and basic features of the number concept: our brain has been built by evolution to entertain this concept.

Largely unique to humans, however, is the capacity to achieve greater integration and coherence amidst those pre-existing cerebral representations. Jean-Pierre Changeux and I have postulated that the many "modular" brain processors that we have inherited from our evolutionary past are supplemented in humans by a global neuronal workspace (Baars 1989; Dehaene et al. 1998; Dehaene and Naccache, 2001), a system of long-distance cortico-cortical connections that breaks modularity and allows the outputs of essentially any cortical processors to be confronted. In particular, we are able to connect our imprecise concepts of time, space and number to categorically defined words and written symbols. An articulate sequence of such states can be composed mentally, thus supporting chains of deductions that draw together the otherwise dispersed representations of the human brain. Many advances in mathematics, indeed, find their origins in cross-domain links and metaphors (Nunez and Lakoff, 2000). The linking of number and space concepts, for instance, leads to capital advances at several points in the history of mathematics, from the invention of numerical measurement to the discovery of irrational numbers, Cartesian coordinates, and algebraic geometry.

Mathematical reality is thus a cultural and mental construction, but one which draws upon the constraints imposed on us by millions of years of brain evolution. Let us examine briefly how this coarse characterization may begin to address the two puzzles of mathematical truth. First, why are mathematical truths universally agreed upon? As noted by Kant, "The science of mathematics presents the most brilliant example of how pure reason may successfully enlarge its domain without the aid of experience." Being independent of the vagaries of experience, mathematics would then reflect the universal structure of our cerebral representations. Thus, I tentatively surmise that, if the products of mathematics appear to us as a rigid body of absolute truths, it is perhaps because they are tightly constrained by the pre-existing structure of our representations. As structural properties of our brains, our representations of space, time, and number are universally shared – as are our modest logical deduction abilities. Once we start from the same intuitive axioms, and follow the same rules of logical deduction, it should not be so surprising that we ultimately converge on the same truths.

Furthermore, we should not forget that, while the final products of mathematics may achieve a certain perfection, their historical development is fuzzy and chaotic. As stated by W.S. Anglin:

"Mathematics is not a careful march down a well-cleared highway, but a journey into a strange wilderness, where the explorers often get lost. Rigour should be a signal to the historian that the maps have been made, and the real explorers have gone elsewhere."

To affirm that arithmetic is a construction of the human mind does not imply any adherence to cultural relativism. Mathematics is not an *arbitrary* product of the mind: it is a tightly constrained construction that, therefore, shows considerable inter-individual and cross-cultural convergence. Throughout phylogenetic evolution, as well as during cerebral development in childhood, selection acts to ensure that the brain constructs internal representations that are adapted to the external world. For instance, at our scale, the world is mostly made up of separable objects that combine into sets according to the familiar equation 1+1=2. I believe that this might be why evolution has anchored this and other rules of arithmetic in our brains.

The notion of an evolution of mathematics provides some insight into the second puzzle of mathematical truth, namely, why mathematics is able to represent the physical world with such remarkable precision. Mathematicians constantly create new mathematical "objects," many of which are not adapted to the external physical world (those are then called "pure mathematics"). Some of these objects, however, are adapted to the natural world because they are founded on basic representations that have proven useful to survival during phylogenetic evolution (e.g., sense of number, space, time). Other mathematical objects are actively selected as topics of research by mathematicians and physicists precisely because of their explanatory adequacy. In this case, as stated by R.L. Wilder:

"There is nothing mysterious, as some have tried to maintain, about the applicability of mathematics. What we get by abstraction from something can be returned."

In the final analysis, then, Wigner's "unreasonable effectiveness" of mathematics might receive an explanation comparable to that of the astonishing efficiency of the hawk's eye or wing, in a world devoid of an intelligent designer (Dawkins 1996). Both would be accounted for by Darwinian evolution, but, in the case of mathematics, a two-stepped evolutionary process: first a phylogenetic evolution that builds fundamental representations of space, time, and number; and second, a cultural evolution that occurs amongst mental representations, at psychological time scales, and should thus properly be called "mental Darwinism" (Changeux 1983; Changeux and Dehaene 1989). In brief, if today's mathematics is efficient, it is perhaps because yesterday's inefficient mathematics has been ruthlessly eliminated and replaced (Dehaene 1997).

The biological bases of elementary arithmetic

In the remainder of this chapter, I will attempt to demonstrate that this scenario for the origins of mathematic truths has some validity, using recent results from cognitive neuroscience studies. In spite of important advances in psychology and brain imaging techniques, and of a few recent forays into the brain processes underlying algebra (Anderson et al. 2004; Qin et al. 2004), it remains exceedingly difficult to examine the cerebral bases of higher mathematical objects. Thus, research in my laboratory has focused on one of the most basic objects, one that lies at the foundations of mathematics: the concept of number. Our recent data suggest that this concept is universally shared and is rooted in a long evolutionary history, but also that is has been refined thanks to cultural inventions whose acquisition can be investigated in both human children and isolated human cultures.

Consider one of the simplest arithmetic abilities: deciding which of two numbers is the largest. Many experiments have now shown that this elementary arithmetic operation is accessible to laboratory animals and even to untrained animals in the wild (Brannon and Terrace 2000; Hauser et al. 2000; Hauser et al. 2003;

McComb et al. 1994; Rumbaugh et al. 1987; Washburn and Rumbaugh 1991). For instance, macaque monkeys spontaneously choose the larger of two sets of food items (Hauser et al. 2000), and lions spontaneously estimate whether their group is more numerous than another group (McComb et al. 1994).

In a particularly impressive laboratory demonstration, macaque monkeys were trained to order a set of cards by the number of objects that they bore (Brannon and Terrace 2000). After training with cards bearing from one to four objects, monkeys generalized their performance to untrained numbers five through nine. This experiment, like many others, included stringent controls for non-numerical parameters such as density, object size, shape, etc. Thus, it can be concluded that the performance of the animals genuinely reflects an elementary competence to perceive numbers.

In most such tasks, animal performance improves as the distance between the two numbers to be compared increases (distance effect), and also improves as the numbers get smaller (size effect). Altogether, these two effects are well captured by stating that comparison performance depends on the ratio of the compared numbers (Weber's law). Thus, number seems to be represented only approximately, on an internal continuum comparable to that used for other perceptual dimensions, such as weight or height.

What is the evidence, however, that the ability to compare sets in animals has anything to do with the human arithmetic ability? A remarkable finding is that, when humans compare two *Arabic* numerals, one can observe distance and size effects similar to those found in animals with non-symbolic stimuli (Dehaene et al. 1990; Moyer and Landauer 1967). Thus, human subjects are slower and make more errors when deciding which of eight and nine is the larger, than when comparing five and nine. Similar effects are observed when comparing two-digit numerals such as 78 and 65: the response time curve shows a continuous distance effect, as would be observed if those numbers were presented as sets of objects. One could have thought that numbers presented in symbolic format would be compared with digital precision, by an exact algorithm. However, the tight parallel between comparisons of symbolic and non-symbolic numbers (Buckley and Gillman 1974) rather suggests that humans continue to use an analog representation, comparable to that available to non-human primates, even when processing Arabic symbols. I have suggested that the quantity representation serves as a "core quantity system" towards which new cultural symbols such as digits and words are quickly translated, and which provides the meaning of those symbols (Dehaene 1997).

Recently, with Philippe Pinel, we have pursued the neural basis of the distance effect during number comparison using functional magnetic resonance imaging (fMRI; Pinel et al. 1999, 2001, 2004). Our results indicate that, during number comparison, the activation of the left and right intraparietal sulci (IPS) shows a tight correlation with the behavioral distance effect: it too varies in inverse relation to the distance between the numbers to be compared. Based on a meta-analysis of many fMRI studies of arithmetic tasks, including comparison, calculation (Chochon et al. 1999), approximation (Dehaene et al. 1999), or even the mere detection of digits (Eger et al. 2003), we have suggested that the bilateral horizontal segment of the IPS (HIPS) may play a particular role in the quantity representation (Dehaene et al. 2003).

Indeed, investigations of brain-lesioned patients indicate that a lesion of this region, at least in the left hemisphere, can cause severe deficits of number comprehension and calculation (acalculia). More recently, this region has also been pinpointed as a possible anatomical locus of impairment in children with dyscalculia, a lifelong impairment in arithmetic that cannot be attributed to global mental retardation or to environmental variables. In premature children with dyscalculia, and in a genetic disease called Turner's syndrome, loss of gray matter in the depth of the IPS has been observed with magnetic resonance imaging (Isaacs et al. 2001; Molko et al. 2003, 2004). These results support the argument that the availability of an intact quantity representation in IPS plays an essential role in guiding arithmetic development in children. The IPS may provide children with a foundational intuition of "number sense:" what is a number and how numerical quantities can be compared or combined. If the IPS is impaired, either genetically or accidentally, then number sense is deteriorated and dyscalculia may ensue.

Further support for this argument is provided by the oft-replicated observation that infants already possess elementary numerical abilities comparable to those of other animal species (for review, see Feigenson et al. 2004). Thus, knowledge of numerical quantities predates by many years the acquisition of number words and number symbols. I suggest that it does not merely precede it but actually plays an active role in making it possible.

Number neurons

How might numerosity be encoded by populations of neurons in the IPS? Although animal models of numerical tasks have been known for many decades, it is only very recently that the neural bases of animal numerical abilities have begun to be investigated. Andreas Nieder and Earl Miller (Nieder et al. 2002; Nieder and Miller 2003, 2004) recorded from single neurons in awake monkeys trained to perform a visual number match-to-sample task. Many neurons were tuned to a preferred numerosity: some neurons responded preferentially to sets of one object, others to two objects, and so on up to five objects (Fig. 1). The tuning was coarse and became increasingly imprecise as numerosity increased. The characteristics of this neural code were exactly as expected from a neural network model that had been proposed to account for the distance effect and other characteristics of numerical processing in adults and infants (Dehaene and Changeux 1993). Most important is the location where the number neurons were recorded. Initially, a large proportion was observed in dorsolateral prefrontal cortex, but more recently another population of neurons with a shorter latency was observed in the parietal lobe (Nieder and Miller 2004; see also Sawamura et al. 2002).

The latter number neurons are located in monkey area VIP, in the depth of the intraparietal sulcus, a location that is a plausible homolog of the human HIPS area, which is active during many number tasks. With Olivier Simon, I performed a detailed fMRI investigation of the relative location of the human activation during calculation, relative to other landmarks of the parietal lobe (Simon et al.

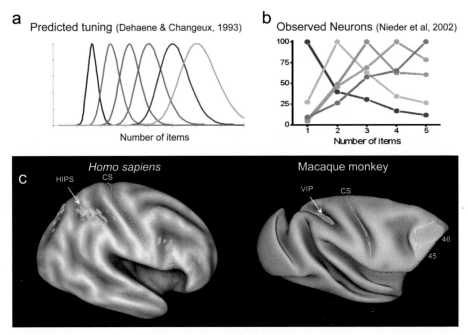

Fig. 1. Number neurons (**a**) as predicted by a neural network model of arithmetic (Dehaene and Changeux 1993) and (**b**) as observed in the macaque monkey prefrontal and parietal cortices (Nieder et al. 2002; Nieder and Miller 2003, 2004). Each neuron responds to a preferred number of visual objects, with a tuning curve that respects Weber's law (the tuning becomes increasingly imprecise for larger numbers). (**c**) Possible homology between number representations in humans and macaque monkeys. The left panel shows an inflated human brain and the location of voxels activated in common to many arithmetic tasks (Dehaene et al. 2003). The right panel shows an inflated monkey brain with the approximate location of prefrontal and parietal areas where a large proportion of number neurons was found.

2002, 2004). This study revealed that number-related activations are anterior to saccade-, attention-, and space-related activations in the posterior IPS (plausibly corresponding to areas LIP and V6a), and posterior to the anterior IPS activations associated with grasping movements (area AIP). Thus, the relative placement of eye movement, calculation and grasping-related brain areas seems to be similar in macaques and humans (Dehaene et al. 2004).

The observation of a common, geometrical brain organization strengthens the possibility that the monkey number neurons in VIP bear a direct evolutionary relation to human arithmetic abilities. However, establishing a genuine interspecies homology would require demonstrating that the human intraparietal cortex also contains distinct populations of neurons, each tuned to a specific number. Yet single neurons cannot be investigated non-invasively in the human brain (although see, e.g., Kreiman et al. 2000). With Manuela Piazza, we recently designed an indirect adaptation method that allowed us to investigate numerosity tuning in humans (Piazza et al. 2004). During fMRI, we repeatedly presented sets of dots with a fixed number (say, 16 dots). The purpose was to "adapt" the neural population coding for this value, thus leading putative human number neurons to progressively reduce their firing rate, as observed in macaque electrophysiological experiments (Miller et al. 1991). We then presented occasional deviant numbers that could range from half to twice the adapted number. fMRI revealed that only two regions, left and right IPS, responded to the change in numerosity by increasing their activation in relation to the distance between the adapted number and the deviant one. Detailed analyses of the activation profile revealed that these regions behave as predicted based on the hypothesis that they contain number neurons. Both human fMRI and monkey electrophysiological data yield tuning profiles that 1) depend only on number independently of other parameters such as shape, density, or spatial arrangement; 2) are smooth and monotonic in response to increasing degrees of deviation in number; 3) are increasingly broader when plotted on a linear scale (Weber's law); and 4) can be expressed as a simple Gaussian function of number ratio. This functional homology, together with the compatible anatomical localization in the depth of the IPS, suggests that humans and macaque monkeys possess similar populations of intraparietal number-sensitive neurons. It provides important support for the notion that all humans start life with a non-verbal representation of approximate number inherited from our evolutionary history

A scenario for the emergence of symbolic arithmetic

One may thus propose a simple scenario for the acquisition of elementary arithmetic in humans. Evolution endowed the primate parietal lobe with a coarse representation of numerosity, which was presumably useful in many situations in which a set of objects or congeners had to be tracked through time. This primitive number representation is also present in humans. It emerges early on in infancy, although its precision is initially quite mediocre and matures during the first year of life (Lipton and Spelke 2003). It provides children with a minimal foundation on which to build arithmetic: the ability to track small sets of objects, to estimate coarse numerosity, and to monitor increases or decreases in numerosity. In the first year, this knowledge is entirely non-verbal, but around three years of age, it becomes connected with symbols, first with the counting words of spoken language (and their surrogate, the fingers), then with the written symbols of the Arabic notation.

An interesting question is whether and how the availability of these cultural symbols affects the arithmetic competence of the human primate. A severe limit of animal arithmetic is that, beyond the numbers 1, 2, 3, arithmetic performance is imprecise: no animal, for instance, has ever shown any ability to precisely discriminate 10 from 11 objects, or to compute 9–8. Humans, on the other hand, are able to perform arithmetic calculations with arbitrary precision (although they remain able to approximate, for instance when verifying a grossly false arithmetic problem such as 13+19 = 92). By allowing reference to discrete numbers in a categorical way, the labels provided by number words and Arabic digits may support a crucial transition from approximate to exact arithmetic.

Recently, my team and I had an opportunity to test this hypothesis directly in a population of humans deprived of verbal labels for large numbers. With Pierre Pica, we studied numerical cognition in the Mundurukú, an Amazonian group whose language includes very few words for numbers (see also Gordon 2004; Pica et al. 2004; see also Gordon 2004). Mundurukú essentially has number words for one through five, plus some quantifiers such as few or many (Fig. 2, top). Using a battery of computerized tasks, we first demonstrated that even those few number words are only used to refer to approximate quantities. When asked to name the numerosity of a set of dots, the Mundurukú used their number words fuzzily, for instance using the word for "four" (ebadipdip) when the actual quantity ranges from three to eight. In spite of this lexical limitation, our participants gave evidence of an excellent understanding of large numbers. They could decide which of two sets of dots was the more numerous, even with numbers ranging up to 80, and even in the presence of considerable variation in non-numerical parameters such as object size or density. They could even perform approximate calculation: when successively shown two sets of objects being hidden in a jar, they could estimate their sum and compare it to a third number. Amazingly, these isolated and non-educated Indians, with a limited language, were as accurate as educated French adults in this non-symbolic approximation task (Fig. 2, middle).

Where they differed, however, is in exact calculation. We presented them with concrete depictions of very simple subtraction problems such as 6-4, by hiding six objects in a jar and then drawing four out. The final result was always 0, 1 or 2, easily within the Mundurukú naming range. In one test, we asked participants to name the result, and in another to point to the correct outcome (zero, one or two objects in the jar). In both cases, the Mundurukú failed to calculate the exact result. They performed relatively well with numbers below three (e. g., 2-2, 3-1), but they failed increasingly frequently as the numbers got larger, not faring better than 50% correct as soon as the initial number exceeded five (Fig. 2, bottom).

We concluded that linguistic labels are not necessary to master the major concepts of arithmetic (quantity, larger-smaller relations, addition, subtraction) and to perform approximate operations. Linguistic coding of numerals, however, may be essential to go beyond this evolutionarily ancient system of approximate arithmetic and to perform exact calculations. If our interpretation is correct, what limits the Mundurukú is not a lack of conceptual knowledge – and thus, our experiments do not provide support for the Whorfian hypothesis that language determines conceptual structure (contra Gordon 2004). Rather, the linguistic coding of numbers is a "cultural tool" that augments the panoply of cognitive strategies available to us to resolve concrete problems. In particular, the mastery of a sequence of number words enables us to count, in a quick and routinize fashion, any number of objects. The Mundurukú do not have a counting routine. Although some possess a rudimentary ability to count on their fingers, it is rarely used. By requiring an exact one-to-one pairing of objects with the sequence of numerals, counting may promote a conceptual integration of approximate number representations, discrete object representations, and the verbal code (Carey 1998; Spelke and Tsivkin 2001). Around the age of three, Western children exhibit an abrupt change in number processing as they suddenly realize that each count word refers to a precise quantity (Wynn 1990). This "crystallization" of discrete numbers out

Mundurukú word	Approximate meaning and range of use
pug ma	one
xep xep	two
ebapug	three (3-5)
ebadipdip	four (3-7)
pug põgbi	five, a handful (5-12)
adesu	some (3-15)
ade	many (7-15)

Approximate calculation
Select which of n1 + n2 or n3 is the larger

(n1) + (n2)

(n3)

Ratio of n1+n2 and n3

— French controls
— All Mundurukú

Exact calculation
Name or point to the result of n1 - n2

(n1)

(n2)

Magnitude of n1

— French controls
— All Mundurukú

Fig. 2. Explorations of arithmetic in an Amazonian culture with few number words. The Mundurukú language only has number words up to five, and these words mostly refer to an approximate range rather than a precise quantity (top). Mundurukú participants were almost as precise as educated French subjects in a task of approximate numerosity addition (middle). However, they failed in an exact calculation task as soon as the numbers exceeded three (bottom).

←——————————————————————————

of an initially approximate continuum of numerical magnitudes does not seem to occur in the Mundurukú.

Conclusion

Within the miniature world of elementary arithmetic, we can track the origins of some mathematical truths in their simplest form. How do we know, with absolute certainty, that statements such as $1+1=2$, $6-4=2$, or $e^{i\pi} = -1$ are true? My research suggests that these statements occupy different steps within a hierarchy of cultural and cerebral constructions. The most basic arithmetic truths are deeply engraved in our brains. After millions of years of evolution in a world made of movable objects, our brains anticipate that one object plus one object should make two objects, and even infants have access to such truths. Others are arrived at by a mental synthesis of two domains. The formula $6-4 = 2$, for instance, does not seem to be accessible or expressible by Mundurukú speakers yet, because it requires an integration of verbal and quantity representations. It is exciting that, using psychological and neuroscientific tools, we can begin to explore the cerebral representations that underlie such knowledge. Whether we will ultimately be able to tell anything useful, at the cerebral level, about $e^{i\pi} = -1$ or any other of the truths of advanced mathematics remains to be seen.

References

Anderson JR, Qin Y, Stenger VA, Carter CS (2004) The relationship of three cortical regions to an information-processing model. J Cogn Neurosci 16: 637–653

Baars BJ (1989) A cognitive theory of consciousness. Cambridge, Mass.: Cambridge University Press

Brannon EM, Terrace HS (2000) Representation of the numerosities 1–9 by rhesus macaques (Macaca mulatta). J Exp Psychol Animal Behav Proc 26: 31–49

Buckley PB, Gillman CB (1974) Comparison of digits and dot patterns. J Exp Psychol 103: 1131–1136

Carey S (1998) Knowledge of number: its evolution and ontogeny. Science 282: 641–642

Changeux JP (1983) L'homme neuronal. Paris: Fayard.

Changeux JP (2002) L'homme de vérité. Paris: Odile Jacob

Changeux JP, Connes A (1989) Matière à pensée. Paris: Odile Jacob

Changeux JP, Dehaene S (1989) Neuronal models of cognitive functions. Cognition 33: 63–109

Chochon F, Cohen L, van de Moortele PF, Dehaene S (1999). Differential contributions of the left and right inferior parietal lobules to number processing. J Cogn Neurosci 11: 617–630

Dawkins R (1996) The blind watchmaker: why the evidence of evolution reveals a universe without design. New York: W.W.Norton

Dehaene S (1997) The number sense. New York: Oxford University Press.

Dehaene S, Changeux JP (1993) Development of elementary numerical abilities: a neuronal model. J Cogn Neurosci 5: 390–407

Dehaene S, Dupoux E, Mehler J (1990) Is numerical comparison digital: analogical and symbolic effects in two-digit number comparison. J Exp Psychol Human Perception Performance 16: 626–641

Dehaene S, Kerszberg M, Changeux JP (1998) A neuronal model of a global workspace in effortful cognitive tasks. Proc Natl Acad Sci USA 95: 14529–14534

Dehaene S, Spelke E, Pinel P, Stanescu R, Tsivkin S (1999) Sources of mathematical thinking: behavioral and brain-imaging evidence. Science 284: 970–974

Dehaene S, Naccache L (2001) Towards a cognitive neuroscience of consciousness: basic evidence and a workspace framework. Cognition 79: 1–37

Dehaene S, Piazza M, Pinel P, Cohen L (2003) Three parietal circuits for number processing. Cogn Neuropsychol 20: 487–506

Dehaene S, Molko N, Cohen L, Wilson AJ (2004) Arithmetic and the brain. Curr Opin Neurobiol 14: 218–224

Eger E, Sterzer P, Russ MO, Giraud AL, Kleinschmidt A (2003) A supramodal number representation in human intraparietal cortex. Neuron 37: 719–725.

Feigenson L, Dehaene S, Spelke E (2004) Core systems of number. Trends Cogn Sci 8: 307–314

Gordon P (2004) Numerical cognition without words: Evidence from Amazonia. Science, 306:496–499

Hauser MD, Carey S, Hauser LB (2000) Spontaneous number representation in semi-free-ranging rhesus monkeys. Proc R Soc Lond B Biol Sci 267: 829–833

Hauser MD, Tsao F, Garcia P, Spelke ES (2003) Evolutionary foundations of number: spontaneous representation of numerical magnitudes by cotton-top tamarins. Proc R Soc Lond B Biol Sci 270: 1441–1446

Isaacs EB, Edmonds CJ, Lucas A, Gadian DG (2001) Calculation difficulties in children of very low birthweight: a neural correlate. Brain 124(Pt 9): 1701–1707

Kreiman G, Koch C, Fried I (2000) Imagery neurons in the human brain. Nature 408: 357–361

Lipton J, Spelke E (2003) Origins of number sense: Large number discrimination in human infants. Psych Sci14: 396–401

McComb K, Packer C, Pusey A (1994) Roaring and numerical assessment in contests between groups of female lions, Panthera leo. Animal Behav 47: 379–387

Miller EK, Li L, Desimone R (1991) A neural mechanism for working and recognition memory in inferior temporal cortex. Science 254: 1377–1379

Molko N, Cachia A, Riviere D, Mangin JF, Bruandet M, Le Bihan D, Cohen L, Dehaene S (2003) Functional and structural alterations of the intraparietal sulcus in a developmental dyscalculia of genetic origin. Neuron 40: 847–858.

Molko N, Cachia A, Riviere D, Mangin JF, Bruandet M, LeBihan D, Cohen L, Dehaene S (2004) Brain anatomy in Turner syndrome: evidence for impaired social and spatial-numerical networks. Cereb Cortex 14:840–850

Moyer, RS, Landauer TK (1967) Time required for judgements of numerical inequality. Nature 215: 1519–1520

Nieder A, Miller EK (2003) Coding of cognitive magnitude. Compressed scaling of numerical information in the primate prefrontal cortex. Neuron 37: 149–157

Nieder A, Miller EK (2004) A parieto-frontal network for visual numerical information in the monkey. Proc Natl Acad Sci USA 101: 7457–7462

Nieder A, Freedman DJ, Miller EK (2002) Representation of the quantity of visual items in the primate prefrontal cortex. Science 297: 1708–1711

Nunez RE, Lakoff G (2000) Where mathematics comes from: how the embodied mind brings mathematics into being. New York: Basic Books

Piazza M, Izard V, Pinel P, Le Bihan D, Dehaene S (2005) Tuning curves for numerosity in the human intraparietal sulcus. Neuron, 44: 547–555

Pica P, Lemer C, Izard V, Dehaene S (2004) Exact and approximate arithmetic in an Amazonian indigene group. Science 306: 499–503

Pinel, P, Le Clec'H G, van de Moortele PF, Naccache L, Le Bihan D, Dehaene S (1999) Event-related fMRI analysis of the cerebral circuit for number comparison. Neuroreport 10: 1473–1479

Pinel P, Dehaene S, Riviere D, LeBihan D (2001) Modulation of parietal activation by semantic distance in a number comparison task. Neuroimage 14: 1013–1026.

Pinel P, Piazza M, Le Bihan D, Dehaene S (2004) Distributed and overlapping cerebral representations of number, size, and luminance during comparative judgments. Neuron 41: 983–993

Qin Y, Carter CS, Silk EM, Stenger VA, Fissell K, Goode A, Anderson JR (2004) The change of the brain activation patterns as children learn algebra equation solving. Proc Natl Acad Sci USA 101: 5686–5691

Rumbaugh DM, Savage-Rumbaugh S, Hegel MT (1987) Summation in the chimpanzee (Pan troglodytes). J Exp Psychol Animal Behav Proc13: 107–115

Sawamura H, Shima K, Tanji J (2002) Numerical representation for action in the parietal cortex of the monkey. Nature 415: 918–922

Simon O, Mangin JF, Cohen L, Le Bihan D, Dehaene S (2002) Topographical layout of hand, eye, calculation, and language-related areas in the human parietal lobe. Neuron 33: 475–487

Simon O., Kherif F, Flandin G, Poline JB, Riviere D, Mangin JF, Le Bihan D, Dehaene S (2004). Automatized clustering and functional geometry of human parietofrontal networks for language, space, and number. Neuroimage 23: 1192–1202

Spelke E, Tsivkin S (2001) Initial knowledge and conceptual change: space and number. In: Bowerman M, Levinson SC (eds) Language acquisition and conceptual development. Cambridge: Cambridge University Press, pp. 70–100

Washburn DA, Rumbaugh DM (1991) Ordinal judgments of numerical symbols by macaques (Macaca mulatta). Psychol Sci 2: 190–193

Wynn K (1990) Children's understanding of counting. Cognition 36: 155–193.

Subject Index

Printing: Krips bv, Meppel
Binding: Stürtz, Würzburg